# Lecture Notes in Mathematics

Edited by J.-M. Morel, F. Takens and B. Teissier

**Editorial Policy**
for the publication of monographs

1. Lecture Notes aim to report new developments in all areas of mathematics – quickly, informally and at a high level. Monograph manuscripts should be reasonably self-contained and rounded off. Thus they may, and often will, present not only results of the author but also related work by other people. They may be based on specialized lecture courses. Furthermore, the manuscripts should provide sufficient motivation, examples and applications. This clearly distinguishes Lecture Notes from journal articles or technical reports which normally are very concise. Articles intended for a journal but too long to be accepted by most journals, usually do not have this "lecture notes" character. For similar reasons it is unusual for doctoral theses to be accepted for the Lecture Notes series.

2. Manuscripts should be submitted (preferably in duplicate) either to one of the series editors or to Springer-Verlag, Heidelberg. In general, manuscripts will be sent out to 2 external referees for evaluation. If a decision cannot yet be reached on the basis of the first 2 reports, further referees may be contacted: the author will be informed of this. A final decision to publish can be made only on the basis of the complete manuscript, however a refereeing process leading to a preliminary decision can be based on a pre-final or incomplete manuscript. The strict minimum amount of material that will be considered should include a detailed outline describing the planned contents of each chapter, a bibliography and several sample chapters.
Authors should be aware that incomplete or insufficiently close to final manuscripts almost always result in longer refereeing times and nevertheless unclear referees' recommendations, making further refereeing of a final draft necessary.
Authors should also be aware that parallel submission of their manuscript to another publisher while under consideration for LNM will in general lead to immediate rejection.

3. Manuscripts should in general be submitted in English.
Final manuscripts should contain at least 100 pages of mathematical text and should include
- a table of contents;
- an informative introduction, with adequate motivation and perhaps some historical remarks: it should be accessible to a reader not intimately familiar with the topic treated;
- a subject index: as a rule this is genuinely helpful for the reader.

Continued on inside back-cover

# Lecture Notes in Mathematics 1756

Editors:
J.-M. Morel, Cachan
F. Takens, Groningen
B. Teissier, Paris

**Springer**

*Berlin
Heidelberg
New York
Barcelona
Hong Kong
London
Milan
Paris
Singapore
Tokyo*

Peter E. Zhidkov

# Korteweg-de Vries and Nonlinear Schröginger Equations: Qualitative Theory

 Springer

Author

Peter E. Zhidkov
Bogoliubov Laboratory of Theoretical Physics
Joint Institute for Nuclear Research
141980 Dubna, Russia

E-mail: zhidkov@thsun1.jinr.ru

Cataloging-in-Publication Data applied for

Die Deutsche Bibliothek - CIP-Einheitsaufnahme

Zhidkov, Peter E.:
Korteweg-de Vries and nonlinear Schrödinger equations : qualitative
theory / Peter E. Zhidkov. - Berlin ; Heidelberg ; New York ;
Barcelona ; Hong Kong ; London ; Milan ; Paris ; Singapore ; Tokyo :
Springer, 2001
  (Lecture notes in mathematics ; 1756)
  ISBN 3-540-41833-4

Mathematics Subject Classification (2000): 34B16, 34B40, 35D05, 35J65,
35Q53, 35Q55, 35P30, 37A05, 37K45

ISSN 0075-8434
ISBN 3-540-41833-4 Springer-Verlag Berlin Heidelberg New York

Springer-Verlag Berlin Heidelberg New York
a member of BertelsmannSpringer Science+Business Media GmbH

http://www.springer.de

© Springer-Verlag Berlin Heidelberg 2001
Printed in Germany

Typesetting: Camera-ready T$_E$X output by the authors
SPIN: 10759936      41/3142-543210 - Printed on acid-free paper

*Dedicated to the memory of my mother*
*Zhidkova Veronika Petrovna*

# Contents

# Introduction

During the last 30 years the theory of solitons – the theory of nonlinear partial differential equations (PDEs) possessing solutions of a special kind – has grown into a large field that attracts the attention of both mathematicians and physicists in view of its important applications and of the novelty of the problems. Physical problems leading to the equations under consideration are observed, for example, in the monograph by V.G. Makhankov [60]. One of the related mathematical discoveries is the possibility of studying certain nonlinear equations from this field by methods that were developed to analyze the quantum inverse scattering problem; these equations are called solvable by the method of the inverse scattering problem (on this subject, see, for example [89,94]). At the same time, the class of currently known nonlinear PDEs solvable by this method is sufficiently narrow and, on the other hand, there is another approach, called the qualitative theory of differential equations. The latter approach in particular includes investigations on the well-posedness of various problems for these equations, the behavior of solutions such as stability or blowing-up, properties of dynamical systems generated by these equations, etc., and this approach makes it possible to investigate an essentially wider class of problems (maybe in a more general study).

In the present book, the author surveys some problems and methods of the qualitative theory of equations under consideration, both stationary and evolutionary, that he has dealt with during about twenty years. So, the selection of the material is mainly related to the author's scientific interests. There are four main topics. These are results on the existence of solutions for initial-value problems for these equations, studies of stationary problems arising when solutions of special kinds (for example, travelling or standing waves) are substituted in the equations under consideration, problems of the stability of solitary waves, and the construction of invariant measures for dynamical systems generated by the Korteweg–de Vries and nonlinear Schrödinger equations.

We consider the following (generalized) Korteweg–de Vries equation (KdVE)

$$u_t + f(u)u_x + u_{xxx} = 0$$

and the nonlinear Schrödinger equation (NLSE)

$$iu_t + \Delta u + f(|u|^2)u = 0,$$

where $i$ is the imaginary unit, $u = u(x,t)$ is an unknown function (real in the first case and complex in the second), $t \in R$, $x \in R$ in the case of the KdVE and $x \in R^N$ for the NLSE with a positive integer $N$, $f(\cdot)$ is a smooth real function and $\Delta = \sum_{k=1}^{N} \frac{\partial^2}{\partial x_k^2}$

is the Laplacian. Typical examples, important for physics, of the functions $f(s)$ and $f(s^2)$ respectively, are the following:

$$|s|^\nu, \quad \frac{as^2}{1+s^2}, \quad e^{-as^2} \quad \text{(where } a \text{ and } \nu \text{ are positive constants).}$$

Chapter 1 contains results on the well-posedness of the Cauchy problem and initial-boundary value problems for the KdVE and the NLSE used further. In Chapter 2, we consider the stationary problem. It arises when we substitute the expression for travelling waves $u = \phi(\omega, x - \omega t)$ in the case of the KdVE and standing waves $u = e^{i\omega t}\phi(\omega, x)$ in the case of the NLSE, where $\omega \in R$ and $\phi$ is a real function, into the equation (it is convenient to introduce just this notation, specifying, if necessary, what equation is being dealt with). In what follows, the solutions of these kinds will be called the *solitary waves* if $\phi$ is a bounded function possessing limits as $x \to \pm\infty$ (as $|x| \to \infty$ for the NLSE). Substituting the expression for standing waves into the NLSE, we obtain the following nonlinear elliptic equation of the second order:

$$\Delta\phi - \omega\phi + f(|\phi|^2)\phi = 0, \quad x \in R^N,$$

which we supplement with some boundary conditions, for example,

$$\phi|_{|x|\to\infty} = 0.$$

A similar problem arises for the KdVE. For the KdVE and the NLSE with $N = 1$, the problem of existence and uniqueness of solitary waves satisfying conditions of the type $\phi^{(k)}(\pm\infty) = 0$ $(k = 0, 1, 2)$ can be sufficiently easily solved (see Chapter 2). Difficulties arise when $N \geq 2$. In this case, generally speaking, non-uniqueness of nontrivial solutions occurs when such solutions exist (for example, for functions $f$ of the above kinds). Let us consider solutions depending only on the argument $r = |x|$. In this case, the typical result for functions $f$ interesting for us is the following: for any integer $l \geq 0$ there exists a solution of our problem which, as a function of the argument $r = |x|$, has exactly $l$ roots on the half-line $r > 0$.

We consider two methods of proving the existence of solitary waves. These are the method of the qualitative theory of ordinary differential equations (ODEs) and the variational method. As an example of the latter, we briefly consider the concentration-compactness method of P.L. Lions. In addition, in this chapter we touch upon recent results on the property of being a basis (for example, in $L_2$) for systems of eigenfunctions of nonlinear one-dimensional Sturm-Liouville-type problems in finite intervals similar to those indicated above.

Chapter 3 is devoted to the stability of solitary waves, which is understood in the Lyapunov sense. Omitting some details, this means that, if for an arbitrary $u_0$ from a set $X$, equipped with a distance $R(\cdot, \cdot)$, there exists a unique solution $u(t)$, $t > 0$, of the equation under consideration, belonging to $X$ for any fixed $t > 0$, then a solution $\Psi(t)$, belonging to $X$ for any fixed $t \geq 0$ is called stable with respect to the distance $R$, if for any $\epsilon > 0$ there exists $\delta > 0$ such that for any solution $u(t)$, belonging to $X$ for any fixed $t \geq 0$ and satisfying $R(\Psi(0), u(0)) < \delta$, one has $R(\Psi(t), u(t)) < \epsilon$ for all $t > 0$. Probably the historically first result on the stability of solitary waves was that obtained by A.N. Kolmogorov, I.G. Petrovskii and N.S. Piskunov [48]: in our terminology, they proved (in particular) a stability of a kink for a nonlinear diffusion equation (in the one-dimensional case a solitary wave is called a *kink* if $\phi'_x(\omega, x) \neq 0$ for all $x$).

Let us introduce a special distance in the real Sobolev space $H^1$ consisting of functions of the argument $x$, by the following rule:

$$\rho(u, v) = \inf_{\tau \in R^N} \|u(\cdot) - v(\cdot + \tau)\|_{H^1}.$$

If we call two functions $u$ and $v$ from $H^1$, satisfying the condition $u(x) \equiv v(x - \tau)$ for some $\tau \in R$, equivalent, then the set of classes of equivalent functions with the distance $\rho$ becomes a metric space. For several reasons, it is natural to investigate the stability of solitary waves of the KdVE with respect to this distance $\rho$; first, because the KdVE is invariant up to translations in $x$; second, the KdVE usually possesses a smooth family of solitary waves depending on the parameter $\omega \in (a, b)$. Therefore any two solutions $\phi(\omega_1, x - \omega_1 t)$ and $\phi(\omega_2, x - \omega_2 t)$, close to each other at the point $t = 0$ in the sense of standard functional spaces such as Lebesgue or Sobolev spaces, cannot be close in the same sense for all $t > 0$ if they have non-equal velocities $\omega_1$ and $\omega_2$. At the same time, if two solitary waves are close at $t = 0$ in the sense of the distance $\rho$, then they can be easily verified to be close for all $t > 0$ in the same sense. T.B. Benjamin in his pioneering paper [7] has proved the stability of solitary waves of the usual KdVE with $f(s) = s$ with respect to the distance $\rho$. He called this stability the *stability of the form* of a solitary wave. Later, his approach was developed by many authors and we shall consider their results.

For solitary waves of the NLSE, the distance $\rho$ should be modified. It should be taken in the following form:

$$d(u, v) = \inf_{\tau, \gamma} \|u(\cdot) - e^{i\gamma} v(\cdot - \tau)\|_{H^1} \quad (u, v \in H^1)$$

where $H^1$ is now the complex space, $\tau \in R^N$ and $\gamma \in R$. To clarify this fact, we only remark here that the usual one-dimensional cubic NLSE with $f(s) = s$ has a two-parameter family of solutions

$$\Phi(x, t) = \sqrt{2\omega} \exp \left\{ i[bx - (b^2 - \omega)t] \right\} \left\{ \cosh[\sqrt{\omega}(x - 2bt)] \right\}^{-1}$$

where $\omega > 0$ and $b$ are real parameters. Therefore, two arbitrary solutions from this family, arbitrary close at $t = 0$ in the sense of the distance $\rho$, cannot be close for all

4

$t > 0$ in the same sense if they correspond to different values of $b^2 - \omega$. By analogy, any two standing waves of the NLSE, close at $t = 0$ in the sense of the distance $\rho$ and corresponding to two values of the parameter $\omega$, nonequal to each other, cannot be close in the sense of the distance $\rho$ for all $t > 0$. At the same time, the functions of the family $\Phi(x, t)$ above satisfy the definition of stability in the sense of the distance $d$.

In the two cases of the KdVE and the NLSE, we present a sufficient (and "almost necessary") condition for the stability of solitary waves satisfying $\lim_{|x| \to \infty} \phi(x) = 0$ and $\phi(x) > 0$, that is called the *Q-criterion* in the physical literature. Roughly speaking, for a nonlinearity of general type a solitary wave is stable (with respect to the distance $\rho$ for the KdVE and to $d$ for the NLSE) if the condition $\frac{d}{d\omega} Q(\phi) > 0$ is satisfied.

Next, we consider the stability of kinks for the KdVE with respect to the distance $\rho$. Among physicists there is a widespread opinion that kinks are always stable. Confirming this point of view, we prove the stability of kinks under assumptions on the function $f$ of general type.

The last part of Chapter 3 is devoted to the stability of solitary waves of the NLSE non-vanishing as $|x| \to \infty$. We present a stability of a new and interesting type. It should be said however that many questions remain open in this direction.

In Chapter 4, we deal with the problem of constructing invariant measures for our equations. These objects have many important applications in the theory of dynamical systems. We concentrate our attention on one such application connected with physics. It is the Fermi-Pasta-Ulam phenomenon which is well-known in the theory of nonlinear waves. Roughly speaking, it means the stability according to Poisson of all trajectories of a dynamical system generated by the corresponding equation. By computer simulations, the Fermi-Pasta-Ulam phenomenon was observed for many "soliton" equations. If we have a bounded invariant measure for our dynamical system, then the Poincaré recurrence theorem explains this phenomenon partially. For the NLSE, we construct an invariant measure associated with the conservation of energy and, for the KdVE in the case when it is solvable by the method of the inverse scattering problem, we present an infinite sequence of invariant measures associated with higher conservation laws.

The author wishes to thank all his colleagues and friends for the useful scientific contacts and discussions with them that have contributed importantly to the appearance of the present book.

# Notation

Unless stated otherwise, the spaces of functions introduced are always real in the case of the KdVE and complex for the NLSE.

Everywhere $C, C_1, C_2, C', C'', ...$ denote positive constants.

$N = 1$ for the KdVE and $N$ is a positive integer for the NLSE.

$x = (x_1, ..., x_N) \in R^N$.

$\Delta = \sum_{i=1}^{N} \frac{\partial^2}{\partial x_i^2}$ is the Laplacian.

$R_+ = [0, +\infty)$.

For a measurable domain $\Omega \subset R^N$ $L_p(\Omega)$ $(p \geq 1)$ is the usual Lebesgue space of functions defined on $\Omega$ with the norm $|u|_{L_p(\Omega)} = \{\int_{\Omega} |u(x)|^p dx\}^{\frac{1}{p}}$.

$L_p = L_p(R^N)$ and $|u|_p = |u|_{L_p(R^N)}$.

$(g, h)_{L_2(\Omega)} = \int_{\Omega} g(x)\overline{h(x)}dx$ for any $g, h \in L_2(\Omega)$.

$(\cdot, \cdot) = (\cdot, \cdot)_{L_2(R^N)}$.

$l_2 = \{a = (a_0, a_1, ..., a_n, ...) : a_n \in R, \ ||a||_{l_2}^2 = \sum_{n=0}^{\infty} a_n^2 < \infty\}, \ (a, b)_{l_2} = \sum_{n=0}^{\infty} a_n b_n$
where $a = (a_0, a_1, ..., a_n, ...), \ b = (b_0, b_1, ..., b_n, ...) \in l_2$.

For an open domain $\Omega \subset R^N$ with a smooth boundary $C_0^{\infty}(\Omega)$ is the space of infinitely differentiable functions of the argument $x \in \Omega$ with compact supports in $\Omega$.

$C_0^{\infty} = C_0^{\infty}(R^N)$.

$D$ is the closure of operator $-\Delta$ with the domain $C_0^{\infty}(\Omega)$ in $L_2(\Omega)$. It is well-known that, if $\Omega$ is a bounded domain or $\Omega = R^N$, then $D$ is a positive self-adjoint operator in $L_2(\Omega)$.

Let $\Omega \subset R^N$ be an open bounded domain with a smooth boundary or let $\Omega = R^N$. Then, the space $H^s(\Omega)$ is the completion, according to Hausdorff, of the set $C_0^{\infty}(\Omega)$ equipped with the norm $||u||_{H^s(\Omega)}^2 = |D^{\frac{s}{2}}u|_{L_2(\Omega)}^2 + |u|_{L_2(\Omega)}^2$ $(s \in R)$ and the

6

corresponding scalar product $(\cdot,\cdot)_{H^s(\Omega)}$. Then, $H^s(\Omega)$ are Hilbert spaces (in fact, they are well-known Sobolev spaces).

$$H^s = H^s(R^N), \quad \|\cdot\|_s = \|\cdot\|_{H^s(R^N)} \text{ and } (\cdot,\cdot)_s = (\cdot,\cdot)_{H^s(R^N)}.$$

For an open set $\Omega \subset R^N$, $C(\Omega)$ is the space of continuous bounded functions defined in $\Omega$ with the norm $|u|_{C(\Omega)} = \sup\limits_{x \in \Omega} |u(x)|$.

$C = C(R^N)$ and $|u|_C = |u|_{C(R^N)}$.

For an arbitrary positive integer $k$, $X^k$ is the Banach space consisting of functions $u(x)$ of the argument $x \in R$, absolutely continuous with their derivatives of order $1, 2, ..., k-1$ in any finite interval, for each of which the following norm is finite:

$$\||u\||_k = |u|_C + \sum_{i=1}^k |\frac{d^i u}{dx^i}|_2.$$

Let $C^k(I; X)$, where $k = 0, 1, 2, ...$, $X$ is a Banach space with a norm $\|\cdot\|$ and $I \subset R$ is a connected set, be the space of $k$ times continuously differentiable functions $u : I \to X$ with derivatives in $t$ of order $0, 1, ..., k$ bounded in $I$, with the norm $\|u\|_{C^k(I;X)} = \max\limits_{k=0,1,...,k} \sup\limits_{t \in I} \|\frac{d^k u(t)}{dt^k}\|$.

$S$ is the Schwartz space consisting of infinitely differentiable functions $u(x)$ of the argument $x \in R$ rapidly decreasing as $|x| \to \infty$ so that for any $k, l = 0, 1, 2, ...$ $p_{k,l}(u) = \sup\limits_x \left|x^l \frac{d^k u(x)}{dx^k}\right| < \infty$, with the topology generated by the system of seminorms $p_{k,l}(\cdot)$.

For an open domain $\Omega \subset R^N$ with a sufficiently smooth boundary, let $W_p^l(\Omega)$, where $p \geq 1$ and $l = 1, 2, 3, ...$, be the standard Sobolev space being the completion of the set $C_0^\infty(\Omega)$ equipped with the norm

$$\|u\|_{W_p^l(\Omega)} = \left\{\int_\Omega \left[|u(x)|^p + \sum_{k_1,...,k_N: \, k_1+...+k_N=l} \left|\frac{\partial^l u(x)}{\partial x_1^{k_1}...\partial x_N^{k_N}}\right|^p\right] dx\right\}^{\frac{1}{p}}.$$

Let $W_p^l = W_p^l(R^N)$. Then, clearly $W_2^l = H^l$, $l = 1, 2, 3, ....$

The space $C^\infty(I; S)$, where $I$ is an interval, is the set of infinitely differentiable functions $u(x,t)$ defined for $(x,t) \in R \times I$, belonging to the space $S$ for any fixed $t \in I$ and such that for any integer $k, l, m \geq 0$

$$\sup\limits_{x \in R, \, t \in I} \left|x^k \frac{\partial^{l+m} u(x,t)}{\partial x^l \partial t^m}\right| < \infty.$$

$\nabla = \left(\frac{\partial}{\partial x_1}, ..., \frac{\partial}{\partial x_N}\right)$ is the gradient.

For an integer $n \geq 0$ and $A > 0$, $H^n_{\text{per}}(A)$ is the completion of the linear space of infinitely differentiable functions defined in $R$ and periodic with the period $A$, equipped with the norm

$$\|u\|_{H^n_{\text{per}}(A)} = \left\{ \int_0^A \left[ u^2(x) + \left(\frac{d^n u(x)}{dx^n}\right)^2 \right] dx \right\}^{\frac{1}{2}} .$$

# Chapter 1

# Evolutionary equations. Results on existence

In this chapter we consider several results on the well-posedness of initial-value problems for the KdVE and NLSE that are used in the next sections. In Additional remarks to this chapter, we mention additional literature on this subject. Now we prove the result, generally well-known, which is intensively exploited further.

Gronwell's lemma. *Let a nonnegative function $y(t)$, defined and continuous on a segment $[t_0, T]$, satisfy the inequality*

$$y(t) \leq a \int_{t_0}^{t} y(s)ds + b, \quad t \in [t_0, T],$$

*where $a$ and $b$ are positive constants. Then,*

$$\int_{t_0}^{t} y(s)ds \leq \frac{b}{a}e^{a(t-t_0)} - \frac{b}{a} \quad and \quad y(t) \leq be^{a(t-t_0)}$$

*for all $t \in [t_0, T]$.*

Proof. Let $Y(t) = \int_{t_0}^{t} y(s)ds$. Then,

$$\frac{\dot{Y}(t)}{aY + b} \leq 1, \quad t \in [t_0, T].$$

Integrating this inequality from $t_0$ to $t$, we get

$$\frac{1}{a} \ln \frac{aY(t) + b}{b} \leq t - t_0, \quad t \in [t_0, T].$$

Hence

$$y(t) \leq aY(t) + b \leq be^{a(t-t_0)}, \quad t \in [t_0, T]$$

and the Gronwell's lemma is proved.□

## 1.1   The (generalized) Korteweg–de Vries equation (KdVE)

In this section we establish two results on the well-posedness of the Cauchy problem for the KdVE

$$u_t + f(u)u_x + u_{xxx} = 0, \quad x, t \in R, \tag{I.1.1}$$

$$u(x,0) = u_0(x). \tag{I.1.2}$$

At first, we consider the case when solutions of this problem vanish as $|x| \to \infty$. We introduce the following definition of generalized solutions in this case.

Definition I.1.1 *Let* $u_0(\cdot) \in H^2$, $f(\cdot) \in \bigcup_{m=1}^{\infty} C((-m,m); R)$ *and* $T_1, T_2 > 0$ *be arbitrary. We call a function* $u(\cdot,t) \in C((-T_1, T_2); H^2) \cap C^1((-T_1, T_2); H^{-1})$ *a generalized solution (or a $H^2$-solution) of the Cauchy problem (I.1.1),(I.1.2) if $u(\cdot,0) = u_0(\cdot)$ in the space $H^2$ and after the substitution of this function in (I.1.1) the equality holds in the sense of the space $H^{-1}$ for any $t \in (-T_1, T_2)$.*

Remark I.1.2 *Since in Definition I.1.1 $u(\cdot,t) \in C((-T_1, T_2); H^2)$, clearly we have $f(u(\cdot,t))u_x(\cdot,t) + u_{xxx} \in C((-T_1, T_2); H^{-1})$. We also note that in view of Definition I.1.1 if $u(\cdot,t)$ is a generalized solution of the problem (I.1.1),(I.1.2) in an interval of time $(-T_1, T_2)$, then it is also a generalized solution of the problem (I.1.1),(I.1.2) in any interval $(-T_1', T_2')$ where $0 < T_i' < T_i$, $i = 1, 2$, so that it is correct to speak about a continuation of a generalized solution onto a wider interval of time. If a $H^2$-solution $u(\cdot,t)$ can be continued on the entire real line $t \in R$ as a function of time so that for any finite interval $I$ containing zero it is a generalized solution in this interval, then we call this solution* global *(defined for all $t \in R$).*

The first result on the well-posedness of the problem (I.1.1),(I.1.2) is the following.

Theorem I.1.3 *Let $f(\cdot)$ be a twice continuously differentiable function satisfying the estimate*

$$|f(u)| \le C(1 + |u|^p) \tag{I.1.3}$$

*with constants $C > 0$ and $p \in (0,4)$ independent of $u \in R$. Then, for any $u_0 \in H^2$ there exists a unique global $H^2$-solution $u(\cdot,t)$ of the problem (I.1.1),(I.1.2). This solution continuously depends on the initial data in the sense that for any $T > 0$ the map $u_0 \longmapsto u(\cdot,t)$ is continuous from $H^2$ into $C((-T,T); H^2) \cap C^1((-T,T); H^{-1})$. In* addition, if $u(\cdot, t)$ is a $H^2$-solution of the problem (I.1.1),(I.1.2), then the quantities

$$E_0(u(\cdot, t)) = \int_{-\infty}^{\infty} u^2(x, t)dx \quad and \quad E_1(u(\cdot, t)) = \int_{-\infty}^{\infty} \left\{ \frac{1}{2}u_x^2(x, t) - F(u(x, t)) \right\} dx,$$

where $\overline{f}(u) = \int_0^u f(s)ds$ and $F(u) = \int_0^u \overline{f}(s)ds$, are determined and independent of $t \in R$, i. e. the functionals $E_0$ and $E_1$ are conservation laws.

A result similar to Theorem I.1.3 takes place for the Cauchy problem (I.1.1),(I.1.2) with periodic initial data. Instead of it we consider a certain result for the problem periodic with respect to the spatial variable $x$ for the standard KdVE with $f(u) = u$; we shall exploit this result in Chapter 4. We introduce the following definition of solutions in the case when the initial data are periodic.

Definition I.1.4 Let $f(u) = u$, $u_0 \in H_{per}^n(A)$ for some $A > 0$ and integer $n \geq 2$ and $T_1, T_2 > 0$. We call a function $u(\cdot, t) \in C((-T_1, T_2); H_{per}^n(A)) \cap C^1((-T_1, T_2); H_{per}^{n-3}(A))$ a solution of the problem (I.1.1),(I.1.2) periodic in $x$ with the period $A > 0$ (or simply a periodic $H^n$-solution) if $u(\cdot, 0) = u_0(\cdot)$ in the space $H_{per}^n(A)$ and, for any $t \in (-T_1, T_2)$, equality (I.1.1) holds in the sense of the space $H_{per}^{n-3}(A)$ after the substitution of the function $u$ in it.

As earlier, it is correct to speak about a continuation of a periodic $H^n$-solution onto a wider interval of time and about a global solution (defined for all $t \in R$). The result on the well-posedness of the problem (I.1.1),(I.1.2) in the periodic case considered in this book is the following.

Theorem I.1.5 Let $f(u) = u$ so that we deal with the standard KdVE. Then for any integer $n \geq 2$ and $u_0 \in H_{per}^n(A)$ there exists a unique global periodic $H^n$-solution of the problem (I.1.1),(I.1.2). This solution continuously depends on the initial data in the sense that for any $T > 0$ the map $u_0 \longmapsto u(\cdot, t)$ is continuous from $H_{per}^n(A)$ into $C((-T, T); H_{per}^n(A)) \cap C^1((-T, T); H_{per}^{n-3}(A))$. In addition there exists a sequence of quantities

$$E_0(u) = \int_0^A u^2(x)dx, \quad E_1(u) = \int_0^A \left\{ \frac{1}{2}u_x^2(x) - \frac{1}{6}u^3(x) \right\} dx,$$

$$E_n(u) = \int_0^A \left\{ \frac{1}{2}[u_x^{(n)}]^2 + c_n u[u_x^{(n-1)}]^2 - q_n(u, ..., u_x^{(n-2)}) \right\} dx, \quad n = 2, 3, 4, ...,$$

where $c_n$ are constants and $q_n$ are polynomials, such that for any integer $n \geq 2$ and a periodic $H^n$-solution $u(\cdot, t)$ of the problem (I.1.1),(I.1.2) (with $f(u) = u$) the quantities $E_0(u(\cdot,t)), ..., E_n(u(\cdot,t))$ do not depend on $t$, i. e. the functionals $E_0, ..., E_n$ are conservation laws for periodic $H^n$-solutions.

Our proof of Theorem I.1.3 consists of several steps. At first, we consider the following regularization of the problem (I.1.1),(I.1.2):

$$w_t + f(w)w_x + w_{xxx} + \epsilon w_x^{(4)} = 0, \quad x \in R, \ t > 0, \ \epsilon > 0, \qquad (I.1.4)$$

$$w(x,0) = w_0(x) \qquad (I.1.5)$$

and prove the following:

<u>Proposition I.1.6</u> *Let $f(\cdot)$ be an infinitely differentiable function satisfying the estimate (I.1.3). Then, for any $u_0 \in S$ and $\epsilon \in (0,1]$ the problem (I.1.4),(I.1.5) has a unique global solution which belongs to $C^\infty([0,n); S)$ for an arbitrary $n = 1,2,3,....$*

At the second step, we take the limit $\epsilon \to +0$ in the problem (I.1.4),(I.1.5). In fact, we get the following statement which is, of course, of an independent interest.

<u>Proposition I.1.7</u> *Let $f(\cdot)$ be an infinitely differentiable function satisfying the estimate (I.1.3). Then, for any $u_0 \in S$ there exists a unique solution $u(\cdot,t) \in \bigcup_{n=1}^\infty C^\infty((-n,n); S)$ of the problem (I.1.1),(I.1.2).*

At the third step, using Proposition I.1.7, we prove Theorem I.1.3.

Now we turn to proving Proposition I.1.6. We begin with the following:

<u>Lemma I.1.8</u> *The system of seminorms*

$$p_{l,0}(u) = \left\{ \int_{-\infty}^{\infty} \left( \frac{d^l u(x)}{dx^l} \right)^2 dx \right\}^{\frac{1}{2}} \quad and \quad p_{0,l}(u) = \left\{ \int_{-\infty}^{\infty} x^{2l} u^2(x) dx \right\}^{\frac{1}{2}}, \quad l = 0,1,2,...$$

*generates the topology in the space $S$.*
<u>Proof</u> follows from the relations

$$p_{m,l}^2(u) = \int_{-\infty}^{\infty} x^{2l} \left( \frac{d^m u(x)}{dx^m} \right)^2 dx = (-1)^m \int_{-\infty}^{\infty} u(x) \frac{d^m}{dx^m} \left[ x^{2l} \frac{d^m u(x)}{dx^m} \right] dx \le$$

$$\le C_{l,m} \sum_{k=0}^{\min\{m;2l\}} \int_{-\infty}^{\infty} \left\{ x^{2(2l-k)} u^2(x) dx + \left( \frac{d^{2m-k}u}{dx^{2m-k}} \right)^2 \right\} dx. \ \square$$

Let us take an arbitrary $\epsilon \in (0,1]$. We construct solutions of the problem (I.1.4),(I.1.5) by the iteration procedure

$$w_{nt} + w_{nxxx} + \epsilon w_{nx}^{(4)} = -f(w_{n-1})w_{(n-1)x}, \quad x \in R, \ t > 0, \ n = 2,3,4,..., \quad \text{(I.1.6)}$$

$$w_n(x,0) = u_0(x), \quad \text{(I.1.7)}$$

where $w_1(x,t;\epsilon) \equiv u_0(x) \in S$. Using the Fourier transform, one can easily show that $w_n \in \bigcup_{m=1}^{\infty} C^{\infty}([0,m); S), \ n = 2,3,4,....$

Taking into account (I.1.3) and applying Sobolev embedding inequalities, we get from (I.1.6):

$$\frac{1}{2}\frac{d}{dt}\int_{-\infty}^{\infty}\left[w_n^2 + \left(\frac{\partial^2 w_n}{\partial x^2}\right)^2\right]dx = -\epsilon\int_{-\infty}^{\infty}\left(\frac{\partial^4 w_n}{\partial x^4}\right)^2 dx - \epsilon\int_{-\infty}^{\infty}\left(\frac{\partial^2 w_n}{\partial x^2}\right)^2 dx -$$

$$-\int_{-\infty}^{\infty}\left(w_n + \frac{\partial^4 w_n}{\partial x^4}\right)f(w_{n-1})\frac{\partial w_{n-1}}{\partial x}dx \le -\epsilon\int_{-\infty}^{\infty}\left\{\left(\frac{\partial^2 w_n}{\partial x^2}\right)^2 + \left(\frac{\partial^4 w_n}{\partial x^4}\right)^2\right\}dx +$$

$$+C_1(\|u_{n-1}\|_2^{p+1} + 1)(|w_{nx}^{(4)}|_2 + |w_n|_2) \le C_2(\epsilon)(1 + \|w_{n-1}\|_2^{2(p+1)} + \|w_n\|_2^2). \quad \text{(I.1.8)}$$

In view of the Gronwell's lemma, inequality (I.1.8) immediately implies the existence of $t_0 = t_0(\epsilon) > 0$ such that

$$\|w_n\|_2^2 \le 2\|u_0\|_2^2, \quad n = 2,3,4,..., \ t \in [0,t_0]. \quad \text{(I.1.9)}$$

Let us now obtain the estimates

$$\left|\frac{\partial^l w_n}{\partial x^l}\right|_2^2 \le c(\epsilon,l), \quad l = 3,4,5,... \quad \text{(I.1.10)}$$

for all $t \in [0,t_0]$ and $n = 2,3,4,....$ By using the induction in $l$, equation (I.1.6), estimate (I.1.9) and embedding theorems, we get:

$$\frac{1}{2}\frac{d}{dt}\left|\frac{\partial^l w_n}{\partial x^l}\right|_2^2 = (-1)^{l+1}\int_{-\infty}^{\infty}\frac{\partial^{2l} w_n}{\partial x^{2l}}(f(w_{n-1})w_{(n-1)x} + w_{nxxx} + \epsilon w_{nx}^{(4)})dx \le -\epsilon\left|\frac{\partial^{l+2} w_n}{\partial x^{l+2}}\right|_2^2 -$$

$$-\int_{-\infty}^{\infty}w_{nx}^{(l+2)}\frac{\partial^{l-2}}{\partial x^{l-2}}\left[f(w_{n-1})w_{(n-1)x}\right]dx \le C(\epsilon,l)$$

and the estimates (I.1.10) are proved.

Now, let us show the existence of $t_1 = t_1(\epsilon) \in (0,t_0(\epsilon)]$ such that for any $m,n = 1,2,3,...$

$$\sup_{0\le t\le t_1}\int_{-\infty}^{\infty}x^{2m}w_n^2 dx \le c_1(m,\epsilon) \quad \text{(I.1.11)}$$

where $c_1(m, \epsilon)$ is a positive constant independent of $n$. We can assume that $m \geq 2$ because the case $m = 1$ can be treated by analogy. By integrating by parts, we derive

$$\frac{1}{2}\frac{d}{dt}\int\limits_{-\infty}^{\infty} x^{2m} w_n^2 dx = -\epsilon \int\limits_{-\infty}^{\infty} x^{2m} w_{nxx}^2 dx + \sum_i P_i$$

where $P_i$ are integrals of the following kinds:

$$P_i = c_i \int\limits_{-\infty}^{\infty} x^{2m-k} w_{nx}^{(r)}\frac{\partial^l w_n}{\partial x^l} dx, \quad P_i = c_i \int\limits_{-\infty}^{\infty} x^{2m-1}\overline{f}(w_{n-1})w_n dx$$

and

$$P_i = c_i \int\limits_{-\infty}^{\infty} x^{2m}\overline{f}(w_{n-1})w_{nx} dx$$

with $k, l = 0, 1, 2$ and $r = 0, 1$. In view of the inequality $x^{2m'} \leq \epsilon_1 x^{2m} + c_2(\epsilon_1, m', m)$, where $0 \leq m' < m$, and the estimates (I.1.9) and (I.1.10), we get the following for the terms $P_i$ of the first type with $l = 2$ and $r = 0$:

$$|P_i| \leq \frac{\epsilon}{K}\int\limits_{-\infty}^{\infty} x^{2m}(w_{nxx})^2 dx + c(\epsilon, m)\int\limits_{-\infty}^{\infty} x^{2m} w_n^2 dx + C_1(\epsilon, m),$$

where $K > 0$ is a sufficiently large constant. The estimates for $l = 0$ or $l = 1$, $r = 0$ are trivial. Consider also the case $r = l = 1$ and $k = 0$ or $k = 1$. Then, we have

$$P_i \leq C_2 + |c_i|\sum_{d=-\infty}^{\infty}\int\limits_{d-1}^{d} x^{2m} w_{nx}^2 dx \leq C_2'\sum_{d=-\infty}^{\infty}\int\limits_{d-1}^{d} x^{2m}\left[c_3(\epsilon, K)w_n^2 + \frac{\epsilon}{K}w_{nxx}^2\right]dx + C_2'' =$$

$$= C_2'\int\limits_{-\infty}^{\infty} x^{2m}\left[c_3(\epsilon, K)w_n^2 + \frac{\epsilon}{K}w_{nxx}^2\right]dx + C_2''$$

where the constant $K > 0$ is arbitrarily large. The terms $P_i$ of other kinds can be estimated by analogy. For example, for the terms $P_i$ of the second kind we have

$$P_i \leq \overline{C} + \overline{\overline{C}}\int\limits_{-\infty}^{\infty} x^{2m}(w_{n-1}^2 + w_n^2)dx.$$

So, we can choose the constant $K > 0$ so large that the term $\epsilon\int_{-\infty}^{\infty} x^{2m} w_{nxx}^2 dx$ becomes larger than the sum of all terms of the kind $\frac{C_2}{K}\int_{-\infty}^{\infty} x^{2m} w_{nxx}^2 dx$. Therefore, we get

$$\frac{1}{2}\frac{d}{dt}\int\limits_{-\infty}^{\infty} x^{2m} w_n^2 dx \leq C(\epsilon, m)\left[1 + \int\limits_{-\infty}^{\infty} x^{2m}(w_{n-1}^2 + w_n^2)dx\right]. \tag{I.1.12}$$

The estimate (I.1.11) follows from (I.1.12).

Inequalities (I.1.9)-(I.1.11) immediately yield the compactness of the sequence $\{w_n\}_{n=1,2,3,\ldots}$ in the space $C([0, t_1(\epsilon)]; S)$. Also, the estimate

$$\frac{1}{2}\frac{d}{dt}\int_{-\infty}^{\infty} g_n^2 dx \le C_3(\epsilon) \int_{-\infty}^{\infty} [g_{n-1}^2 + g_n^2] dx, \quad g_n = w_n - w_{n-1},$$

is implied by equation (I.1.6) and the estimates (I.1.9),(I.1.10). Therefore, the sequence $\{w_n\}_{n=1,2,3,\ldots}$ converges to a function $w(x, t; \epsilon)$ in the space $C([0, t'(\epsilon)]; L_2)$ where $t' \in (0, t_1]$ is sufficiently small. Hence, due to its compactness in $C([0, t_1(\epsilon)]; S)$, this sequence converges to $w(x, t; \epsilon)$ in the space $C([0, t'(\epsilon)]; S)$. Thus, taking the limit in (I.1.6),(I.1.7) as $n \to \infty$, we get the local solvability of the problem (I.1.4),(I.1.5) in the space $C([0, t'(\epsilon)]; S)$.

To show the uniqueness of this solution, let us suppose the existence of two solutions $w^1(x, t; \epsilon)$ and $w^2(x, t; \epsilon)$ of the class $C([0, T]; S)$ with some $T > 0$. Setting $w = w^1 - w^2$, we easily derive from equation (I.1.4):

$$\frac{d}{dt}\int_{-\infty}^{\infty} [w(x, t; \epsilon)]^2 dx \le C(\epsilon) \int_{-\infty}^{\infty} [w(x, t; \epsilon)]^2 dx,$$

where a constant $C(\epsilon) > 0$ does not depend on $x \in R$ and $t \in [0, T]$. Therefore, $w^1 \equiv w^2$ according to the Gronwell's lemma, and the uniqueness of a solution of the above class of the problem (I.1.4),(I.1.5) is proved.

Now we want to make some estimates, uniform with respect to $\epsilon \in (0, 1]$, for solutions of the problem (I.1.4),(I.1.5) of the class $C([0, T]; S)$.

**Lemma I.1.9** *Let the assumptions of Proposition I.1.6 be valid and let $w(x, t; \epsilon) \in C([0, T]; S)$ be a solution of the problem (I.1.4),(I.1.5). Then $|w(\cdot, t; \epsilon)|_2$ is a nonincreasing function of the argument $t > 0$. Also, for any $C > 0$, $p \in (0, 4)$, $R_1 > 0$ and $T > 0$ there exists $R_2 > 0$ such that for an arbitrary infinitely differentiable function $f(\cdot)$, satisfying condition (I.1.3) with these constants $C$ and $p$, any $\epsilon \in (0, 1]$ and a solution $w(x, t; \epsilon) \in C([0, T']; S)$ (where $T' \in (0, T]$ is arbitrary) of the problem (I.1.4),(I.1.5), satisfying the condition $\|w(\cdot, 0; \epsilon)\|_1 \le R_1$, one has $\|w(\cdot, t; \epsilon)\|_1 \le R_2$ for all $t \in [0, T']$.*

Proof. To prove the first statement of our Lemma, it suffices to observe that

$$\frac{1}{2}\frac{d}{dt}\int_{-\infty}^{\infty} w^2(x, t; \epsilon) dx = -\epsilon \int_{-\infty}^{\infty} w_{xx}^2(x, t; \epsilon) dx \le 0.$$

Let us prove the second statement. For a solution $w(x, t; \epsilon) \in C([0, T]; S)$ of the problem (I.1.4),(I.1.5), by applying embedding theorems and the proved statement of the lemma, we get:

$$\frac{1}{2}\frac{d}{dt}\int\limits_{-\infty}^{\infty}w_x^2dx = -\epsilon\int\limits_{-\infty}^{\infty}w_{xxx}^2dx + \frac{d}{dt}\int\limits_{-\infty}^{\infty}F(w(x,t;\epsilon))dx - \epsilon\int\limits_{-\infty}^{\infty}\frac{\partial}{\partial x}[\overline{f}(w)]w_{xxx}dx \le$$

$$\le -\epsilon\int\limits_{-\infty}^{\infty}w_{xxx}^2dx + CC_1\epsilon(1 + |w_{xxx}|_2^{1+\frac{1}{3}+\frac{p}{6}}) + \frac{d}{dt}\int\limits_{-\infty}^{\infty}F(w(x,t;\epsilon))dx \le$$

$$\le \frac{d}{dt}\int\limits_{-\infty}^{\infty}F(w(x,t;\epsilon))dx + C_2\epsilon \qquad (\mathrm{I.1.13})$$

because $1 + \frac{1}{3} + \frac{p}{6} < 2$, and where constants $C_1, C_2 > 0$ depend only on $C, p, R_1$ and on constants $C', C'' > 0$ from the multiplicative inequalities

$$|u|_C \le C'|u|_2^{\frac{5}{6}}|u_{xxx}|_2^{\frac{1}{6}} \quad \text{and} \quad |u_x|_2 \le C''|u|_2^{\frac{2}{3}}|u_{xxx}|_2^{\frac{1}{3}}.$$

Since due to condition (I.1.3) $F(w) \le C_3(u^2 + |u|^{p+2})$ where $p \in (0,4)$, we have by embedding theorems

$$\int\limits_{-\infty}^{\infty}F(w(x,t;\epsilon))dx \le C_3|u|_2^2 + C_4|u|_2^{\frac{p}{2}+2}|u_x|_2^{\frac{p}{2}} \le \frac{1}{4}|u_x|_2^2 + C_5, \qquad (\mathrm{I.1.14})$$

where the following inequality has been used:

$$|u|_{p+2} \le C_6|u|_2^{\frac{1}{2}+\frac{1}{p+2}}|u_x|_2^{\frac{1}{2}-\frac{1}{p+2}} \qquad (\mathrm{I.1.15})$$

and where $C_3, C_4, C_5$ and $C_6$ are positive constants depending only on $R_1, C$ and $p$. Now the second statement of Lemma I.1.9 follows from the first statement, (I.1.13) and (I.1.14).□

  **Lemma I.1.10** *Let $C > 0$, $p \in (0,4)$, $R_1 > 0$ and $T > 0$ be arbitrary and let $R_2 = R_2(C, p, R_1, T) > 0$ be the corresponding constant from Lemma I.1.9. For an arbitrary twice continuously differentiable function $f(\cdot)$ we set:*

$$\overline{R} = \sup_{u \in H^1:\ \|u\|_1 \le R_2}|u|_c, \quad F_2(C, p, f, R_1, T) = \sup_{|u| \le \overline{R}}|f'(u)|$$

*and*

$$F_3(C, p, f, R_1, T) = \sup_{|u| \le \overline{R}}|f''(u)|$$

*(here $\overline{R} < \infty$ in view of the embedding of $H^1$ into $C$). Take an arbitrary sufficiently large $R_3 > 0$. Then, there exists $R_4 > 0$ such that for any $\epsilon \in (0,1]$, an arbitrary*

infinitely differentiable function $f(\cdot)$, satisfying (I.1.3) with the above constants $C$ and $p$ and such that $F_2(C, p, f, R_1, T) \leq R_3$ and $F_3(C, p, f, R_1, T) \leq R_3$, and for an arbitrary solution $w(x, t; \epsilon) \in C([0, T']; S)$ of the problem (I.1.4),(I.1.5) ($T' \in (0, T]$), obeying the conditions $\|w(\cdot, 0; \epsilon)\|_1 \leq R_1$ and $\|w(\cdot, 0; \epsilon)\|_2 \leq R_3$, one has $\|w(\cdot, t; \epsilon)\|_2 \leq R_4$ for all $t \in [0, T']$.

Proof. Take arbitrary constants $R_1$, sufficiently large $R_3$, $C > 0, p \in (0, 4)$, some $\epsilon \in (0, 1]$ and let an infinitely differentiable function $f(\cdot)$ satisfy condition (I.1.3) with these constants $C$ and $p$, $F_2(C, p, f, R_1, T) \leq R_3$ and $F_3(C, p, f, R_1, T) \leq R_3$. Using Lemma I.1.9 and inequality (I.1.15), we get

$$\frac{1}{2}\frac{d}{dt}\int\limits_{-\infty}^{\infty}\left(\frac{\partial^2 w}{\partial x^2}\right)^2 dx = -\epsilon\int\limits_{-\infty}^{\infty}(u_x^{(4)})^2 dx - \int\limits_{-\infty}^{\infty} w_{xx}\frac{\partial^2}{\partial x^2}[f(w)w_x]dx =$$

$$= -\epsilon\int\limits_{-\infty}^{\infty}(w_x^{(4)})^2 dx - \int\limits_{-\infty}^{\infty} f''(w)w_x^3 w_{xx}dx - \frac{5}{2}\int\limits_{-\infty}^{\infty} f'(w)w_x w_{xx}^2 dx =$$

$$= -\epsilon\int\limits_{-\infty}^{\infty}(w_x^{(4)})^2 dx + I_1(w) + I_2(w). \tag{I.1.16}$$

Let us estimate the terms $I_1(w)$ and $I_2(w)$ separately. For $I_1(w)$, applying inequality (I.1.15) and Lemma I.1.9, we get

$$I_1(w) \leq F_3|w_x|_6^3\,|w_{xx}|_2 \leq C_1 F_3|w_x|_2^3\,|w_{xx}|_2^2 \leq C_1 R_2^2 F_3|w_{xx}|_2^2, \tag{I.1.17}$$

where the constant $C_1 > 0$ depends only on the constant from the embedding inequality (I.1.15).

Let us estimate $I_2(w)$. We have

$$I_2(w) = -\frac{5}{6}\frac{d}{dt}\int\limits_{-\infty}^{\infty} f(w)w_{xx}dx + \frac{5}{6}\int\limits_{-\infty}^{\infty}(f(w)f'(w)w_x^3 + f''(w)w_x^3 w_{xx})dx+$$

$$+\frac{5}{6}\epsilon\int\limits_{-\infty}^{\infty}[2f(w)w_{xxx}^2 + f''(w)w_x^3 w_{xxx}dx + 4f'(w)w_x w_{xx}w_{xxx}]dx. \tag{I.1.18}$$

The second term in the right-hand side of this equality can be estimated completely as $I_1(w)$ from (I.1.16), so that we have

$$\frac{5}{6}\int\limits_{-\infty}^{\infty}\{f(w)f'(w)w_x^3 + f''(w)w_x^3 w_{xx}\}dx \leq C_2(F_1 F_2 + F_3)(|w_{xx}|_2^2 + 1) \tag{I.1.19}$$

where $F_1 = \sup\limits_{|u|\leq \overline{R}}|f(u)|$.

Due to embedding theorems, the term from the right-hand side of (I.1.18) with the coefficient $\frac{5}{6}\epsilon$ can be estimated as

$$\epsilon \int\limits_{-\infty}^{\infty} (w_x^{(4)})^2 dx + C_3(F_1, F_2, F_3, R_2). \qquad (I.1.20)$$

Finally, for $I_3(w) = -\frac{5}{6} \int\limits_{-\infty}^{\infty} \bar{f}(w)w_{xx}dx$ we have

$$I_3(w) \le C(1 + C_4\|w\|_1^p)|w|_2 \, |w_{xx}|_2 \le 4C^2 R_2^2 (1 + C_4 R_2^p)^2 + \frac{1}{4}|w_{xx}|_2^2, \qquad (I.1.21)$$

where the constant $C_4 > 0$ depends only on constants from embedding inequalities. In view of Lemma I.1.9 and the estimates (I.1.16)-(I.1.21), Lemma I.1.10 is proved.□

**Lemma I.1.11** *Under the assumptions of Theorem I.1.3 for any integer $l \ge 2$ and $T > 0$ there exists $c(l,T) > 0$ such that for any $\epsilon \in (0,1]$ and an arbitrary solution $w(x,t;\epsilon) \in C([0,T']; S)$ (here $T' \in (0,T]$ is arbitrary) of the problem (I.1.4),(I.1.5) one has $\left|\frac{\partial^l w}{\partial x^l}\right|_2 \le c(l,T)$ for all $t \in [0,T']$.*

Proof. We use the induction in $l$. For $l = 2$ the statement of Lemma is already proved with Lemma I.1.10. Let this statement be valid for $l = 2, ..., r$. Consider the case $l = r + 1$. Using the integration by parts and embedding theorems, we get

$$\frac{1}{2}\frac{d}{dt} \int\limits_{-\infty}^{\infty} \left(\frac{\partial^{r+1} w}{\partial x^{r+1}}\right)^2 dx = -\epsilon \int\limits_{-\infty}^{\infty} \left(\frac{\partial^{r+3} w}{\partial x^{r+3}}\right)^2 dx - \int\limits_{-\infty}^{\infty} \frac{\partial^{r+1}}{\partial x^{r+1}}[f(w)w_x]\frac{\partial^{r+1} w}{\partial x^{r+1}}dx \le$$

$$\le C_1(\|w\|_2) + C_2(\|w\|_2)\left|\frac{\partial^{r+1} w}{\partial x^{r+1}}\right|_2^2 .□$$

**Lemma I.1.12** *Let the assumptions of Theorem I.1.3 be valid and let $T > 0$ and integer $m > 0$ be arbitrary. Then, there exists $c(m) > 0$ such that for any $\epsilon \in (0,1]$ and a solution $w(x,t;\epsilon) \in C([0,T']; S)$, where $T' \in (0,T]$ is arbitrary, of the problem (I.1.4),(I.1.5) the following estimate takes place:*

$$\int\limits_{-\infty}^{\infty} x^{2m} w^2 dx \le c(m), \quad t \in [0,T'].$$

Proof. First of all, we shall show that for any $m = 1, 2, 3, ...$ there exist $C > 0$ and integer $r > 0$ such that for $u \in S$ we have

$$\int\limits_{-\infty}^{\infty} |x|^{2m-l}\left(\frac{d^n u}{dx^n}\right)^2 dx \le C\left\{\|u\|_r^2 + \|u\|_2^2 + \int\limits_{-\infty}^{\infty} x^{2m} u^2(x)dx\right\}, \qquad (I.1.22)$$

where $l = 0$ or $l = 1$ or $l = 2$ and $n = 1$ or $n = 2$. For this aim, we use the obvious estimate

$$\frac{1}{2} \leq \frac{|k|}{|k+x|} \leq 2, \quad x \in (0,1), \quad k = -2, -3, -4, \dots \text{ or } k = 1, 2, 3, \dots \qquad (I.1.23)$$

and the multiplicative inequality

$$\left\{ \int_a^{a+1} \left( \frac{d^n u}{dx^n} \right)^2 dx \right\}^{\frac{1}{2}} \leq C(r) |u|_{L_2(a,a+1)}^{1-\frac{n}{r}} \left( |u_x^{(r)}|_{L_2(a,a+1)}^{\frac{n}{r}} + |u|_{L_2(a,a+1)}^{\frac{n}{r}} \right), \qquad (I.1.24)$$

where $a = 0, \pm 1, \pm 2, \dots$ is arbitrary, $n = 1$ or $n = 2$ and $r > 2$ is arbitrary integer. Due to (I.1.23) and (I.1.24), we get for integer $r > 2mnl^{-1}$:

$$\int_{-\infty}^{\infty} |x|^{2m-l}(u_x^{(n)})^2 dx = \int_{-1}^{1} |x|^{2m-l}(u_x^{(n)})^2 dx + \left( \sum_{k=-\infty}^{-2} + \sum_{k=1}^{\infty} \right) \int_k^{k+1} |x|^{2m-l}(u_x^{(n)})^2 dx \leq$$

$$\leq C'\|u\|_2^2 + C''(r) \left( \sum_{k=-\infty}^{-2} + \sum_{k=1}^{\infty} \right) |k|^{2m-l} \left( \int_k^{k+1} u^2 dx \right)^{1-\frac{n}{r}} \times$$

$$\times \left( |u|_{L_2(k,k+1)} + |u_x^{(r)}|_{L_2(k,k+1)} \right)^{\frac{n}{r}} \leq C'\|u\|_2^2 + C''(r) 2^{2m-l} \times$$

$$\times \left( \sum_{k=-\infty}^{-2} + \sum_{k=1}^{\infty} \right) \left( \int_k^{k+1} x^{2m} u^2 dx \right)^{1-\frac{n}{r}} \left( |u|_{L_2(k,k+1)} + |u_x^{(r)}|_{L_2(k,k+1)} \right)^{\frac{n}{r}} \leq .$$

$$\leq C(r) \left( \|u\|_r^2 + \|u\|_2^2 + \int_{-\infty}^{\infty} x^{2m} u^2 dx \right),$$

where we have used the trivial inequality $|k|^{2m-l} \leq 2^{2m-l} x^{2m-l}$ for $k = -2, -3, \dots$ and $k = 1, 2, 3, \dots$ and $x \in (k, k+1)$, and (I.1.22) follows.

Consider the expression

$$\frac{1}{2}\frac{d}{dt} \int_{-\infty}^{\infty} x^{2m} w^2 dx = -\int_{-\infty}^{\infty} x^{2m} w f(w) w_x dx + 2m \int_{-\infty}^{\infty} x^{2m-1} w w_{xx} dx -$$

$$-m \int_{-\infty}^{\infty} x^{2m-1} w_x^2 dx - \epsilon \int_{-\infty}^{\infty} x^{2m} w w_x^{(4)} dx.$$

Due to Lemmas I.1.9-I.1.11, the Hölder's inequality and (I.1.22), the first, second and third terms in the right-hand side of this equality can be obviously estimated as

$$C_1 + C_2 \int_{-\infty}^{\infty} x^{2m} w^2 dx$$

with some constants $C_1, C_2 > 0$. The last term can be estimated by analogy after an integration by parts. So, we come to the estimate

$$\frac{1}{2}\frac{d}{dt}\int\limits_{-\infty}^{\infty} x^{2m}w^2\,dx \leq C_3 + C_4\int\limits_{-\infty}^{\infty} x^{2m}w^2\,dx.$$

Thus, the statement of Lemma I.1.12 is proved. $\square$

Lemmas I.1.1 and I.1.9-I.1.12 immediately imply the global, for all $t > 0$, solvability of the problem (I.1.4),(I.1.5). Indeed, let $u_0 \in S$. Suppose the existence of $T^* > 0$ such that the corresponding $S$-solution $w(x,t;\epsilon)$ of the problem (I.1.4),(I.1.5), whose uniqueness has already been proved, can be continued onto the half-interval of time $[0, T^*)$ and cannot be continued on an arbitrary right half-neighborhood of the point $t = T^*$. Then, due to the above-indicated results, there exists a limit $\lim\limits_{t\to T^*-0} w(x,t;\epsilon) = u_1 \in S$ understood in the sense of the space $S$. Thus, considering the Cauchy problem for equation (I.1.4) with the initial data $w(x,T^*;\epsilon) = u_1(x)$, we get the local solvability of this problem on an interval of time $[T^*, T^* + \delta)$ with some $\delta > 0$, i. e. we get a contradiction. So, Proposition I.1.6 is proved. $\square$

Let us now prove Proposition I.1.7. Due to Lemmas I.1.9-I.1.12 for any $T > 0$ the existence of a solution $u(x,t) \in C([0,T); S)$ of the problem (I.1.1),(I.1.2) can be obtained by taking the limit as $\epsilon \to +0$ in the problem (I.1.4),(I.1.5). The uniqueness of this solution of the class $C([0,T); S)$ for any $T > 0$ can be proved in the same way as for the problem (I.1.4),(I.1.5). The existence and uniqueness of a solution belonging to $S$ for any fixed $t$ in the domain $t < 0$ can be proved by analogy with the above construction. Thus, Proposition I.1.7 is proved, too. $\square$

Now, we turn to proving Theorem I.1.3. Let us take an arbitrary twice continuously differentiable function $f(\cdot)$ satisfying the estimate (I.1.3) and let $\{f_n(\cdot)\}_{n=1,2,3,...}$ be a sequence of infinitely differentiable functions satisfying the estimate (I.1.3) with the same constants $C$ and $p$ and converging to $f(\cdot)$ in $C^2((-m,m) \times (-l,l); R)$ for any $l, m = 1, 2, 3, ....$ Let us also take arbitrary $u_0 \in H^2$ and $T > 0$ and let $\{u_0^n\}_{n=1,2,3,...} \subset S$ be a sequence converging to $u_0$ weakly in $H^2$ and strongly in $H^1$ as $n \to \infty$. For each $n = 1, 2, 3, ...$ by $u_n(x,t) \in C^\infty((-T,T); S)$ we denote the solution of the problem (I.1.1),(I.1.2) taken with $f = f_n$ and $u_0 = u_0^n$. It is clear that the sequence $\{R_2(C, p, ||u_n||_1, T)\}_{n=1,2,3,...}$, where the function $R_2 > 0$ is given by Lemma I.1.9, is bounded and let $\overline{R}_2 = \sup\limits_n R_2(C, p, ||u_n||_1, T) > 0$. Let also $\overline{R}_3 = \sup\limits_n ||u_0^n||_2$. Then, clearly $\overline{R}_3 \in (0, \infty)$. We set $\overline{R}_4 = R_4(\overline{R}_3)$ where the function $R_4 = R_4(R_3) > 0$

is given by Lemma I.1.10. Then, due to Lemmas I.1.9 and I.1.10,

$$||u_n(\cdot,t)||_1 \leq \overline{R}_2 \quad \text{and} \quad ||u_n(\cdot,t)||_2 \leq \overline{R}_4, \quad t \in (-T,T). \qquad \text{(I.1.25)}$$

For $t \in [-T,0)$ these estimates can be obtained by the simple change of variables $x \to -x$ and $t \to -t$ in equation (I.1.1). Therefore, we have for $t > 0$

$$\frac{1}{2}\frac{d}{dt}\int\limits_{-\infty}^{\infty}(u_n - u_m)^2 dx = -\int\limits_{-\infty}^{\infty}(u_n - u_m)(f_n(u_n)u_{nx} - f_m(u_m)u_{mx})dx =$$

$$= -\int\limits_{-\infty}^{\infty}\{(u_n - u_m)[f(u_n)(u_{nx} - u_{mx}) + u_{mx}(f(u_n) - f(u_m)) +$$

$$+ (f_n(u_n) - f(u_n))u_{nx} + u_{mx}(f(u_m) - f_m(u_m)]\}dx =$$

$$= C(T)\int\limits_{-\infty}^{\infty}(u_n - u_m)^2 dx + a_{n,m},$$

where $a_{n,m} \to +0$ as $n,m \to +\infty$ and by analogy for $t < 0$. These estimates yield the convergence of the sequence $\{u_n\}_{n=1,2,3,\ldots}$ in the space $C([-T,T];L_2)$ to some $u(x,t)$. Due to the estimates (I.1.25),

$$u(\cdot,t) \in H^2 \quad \text{and} \quad ||u(\cdot,t)||_2 \leq \overline{R}_4, \quad t \in [-T,T]. \qquad \text{(I.1.26)}$$

Indeed, let us take an arbitrary $t \in [-T,T]$. Due to (I.1.25), the sequence $\{u_n(\cdot,t)\}_{n=1,2,3,\ldots}$ is weakly compact in $H^2$, hence it contains a weakly converging subsequence (without the loss of the generality we accept that it is the sequence $\{u_n(\cdot,t)\}_{n=1,2,3,\ldots}$). Therefore,

$$u(\cdot,t) \in H^2 \quad \text{and} \quad ||u(\cdot,t)||_2 \leq \liminf_{n\to\infty}||u_n(\cdot,t)||_2 \leq \overline{R}_4$$

and the properties (I.1.26) are proved.

The following statement can be proved by analogy.

<u>Lemma I.1.13</u> *For any $T > 0$ $u_n(\cdot,t) \to u(\cdot,t)$ as $n \to \infty$ in $C((-T,T);H^1)$.*

<u>Lemma I.1.14</u> *If $u_0^n \to u_0$ strongly in $H^2$ as $n \to \infty$, then $||u(\cdot,t)-u_n(\cdot,t)||_2 \to 0$ for any $t \in [-T,T]$ as $n \to \infty$.*

<u>Proof.</u> Due to Lemma I.1.13 and the above arguments $u_n(\cdot,t) \to u(\cdot,t)$ as $n \to \infty$ weakly in $H^2$ for any $t \in R$. Further, we have from (I.1.16),(I.1.18) with $\epsilon = 0$

$$\frac{1}{2}|u_{nxx}(\cdot,t)|_2^2 - \frac{1}{2}|u_{nxx}(\cdot,0)|_2^2 = \int\limits_0^t\int\limits_{-\infty}^{\infty}\{-\frac{1}{6}f_n''(u_n(\cdot,s))u_{nx}^3(\cdot,s)u_{nxx}(\cdot,s) +$$

$$+\frac{5}{6}f_n(u_n(\cdot,s))f'_n(u_n(\cdot,s))u^3_{nx}(\cdot,s)\}dxds + \frac{5}{6}\int\limits_{-\infty}^{\infty}\{f_n(u_n(\cdot,t))u^2_{nx}(\cdot,t)-$$

$$-f_n(u_n(\cdot,0))u^2_{nx}(\cdot,0)\}dx. \tag{I.1.27}$$

We want to show that the right-hand side $R(f_n,u_n)$ in (I.1.27) tends to $R(f,u)$ as $n\to\infty$. Obviously, for the last term in (I.1.27) the passage to the limit is valid. Consider the expression

$$\int\limits_0^t\int\limits_{-\infty}^{\infty}\{f''_n(u_n)u^3_{nx}u_{nxx} - f''(u)u^3_x u_{xx}\}\,dxds =$$

$$= \int\limits_0^t\int\limits_{-\infty}^{\infty}\{[f''_n(u_n) - f''(u)]u^3_{nx}u_{nxx} + f''(u)[u_{nxx}(u^3_{nx} - u^3_x) + u^3_x(u_{nxx} - u_{xx})]\}dxds.$$

Due to embedding theorems, the boundedness of the sequence $\{u_n\}_{n=1,2,3,...}$ in $C((-T,T)\cdot H^2)$, its strong convergence in $H^1$ to $u(\cdot,t)$ following from Lemma I.1.13 and its weak convergence to $u(\cdot,t)$ in $H^2$, the expression in the right-hand side of this equality tends to zero as $n\to\infty$. The other term in the right-hand side of (I.1.27) can be considered by analogy. In addition, we observe that in view of embedding inequalities the absolute value of the integrand of the integral over $(0,t)$ in (I.1.27) is bounded by a positive constant independent of $n$ and $s$. Therefore, indeed $R(f_n,u_n)\to R(f,u)$ as $n\to\infty$.

Let $u_n(\cdot,0)\to u(\cdot,0)$ strongly in $H^2$ as $n\to\infty$. Then, we have

$$\frac{1}{2}|u_{xx}(\cdot,t)|^2_2 \le \liminf_{n\to\infty}|u_{nxx}(\cdot,t)|^2_2 = \frac{1}{2}\liminf_{n\to\infty}|u_{nxx}(\cdot,0)|^2_2 + R(f,u) =$$

$$= \frac{1}{2}|u_{xx}(\cdot,0)|^2_2 + R(f,u). \tag{I.1.28}$$

Now, we observe that all the above considerations are also valid if we change the problem (I.1.1),(I.1.2) by the following:

$$w_t + f(w)w_x + w_{xxx} = 0, \quad w(\cdot,t) = w^0 \in S.$$

Take an arbitrary sequence $\{w_0^n\}_{n=1,2,3,...} \subset S$ converging to $u(\cdot,t)$ weakly in $H^2$. Then, we get that the sequence of infinitely differentiable solutions $w_n(\cdot,s)$ of the above problem, taken with $f = f_n$ and $w^0 = w_0^n$, converges to $u(\cdot,s)$ strongly in $C((-T,T);H^1)$ and, for any fixed $s \in R$, converges weakly in $H^2$ to $u(\cdot,s)$. Let in addition $w_0^n(\cdot)\to u(\cdot,t)$ strongly in $H^2$. Then, we get as earlier

$$\frac{1}{2}|u_{xx}(\cdot,0)|^2_2 \le \frac{1}{2}\liminf_{n\to\infty}|w_{nxx}(\cdot,0)|^2_2 = \frac{1}{2}\liminf_{n\to\infty}|u_{nxx}(\cdot,t)|^2_2 - R(f,u) =$$

$$= \frac{1}{2}|u_{xx}(\cdot,t)|_2^2 - R(f,u).$$

This inequality together with (I.1.28) yields the equality $\lim\limits_{n\to\infty} |u_{nxx}(\cdot,t)|_2 = |u_{xx}(\cdot,t)|_2$ and Lemma I.1.14 is proved. $\square$

**Lemma I.1.15** $u(\cdot,t) \in C((-T,T); H^2)$ *for any* $T > 0$.

**Proof.** Let $u_n(\cdot,0) \to u(\cdot,0)$ strongly in $H^2$ as $n \to \infty$. Taking the limit as $n \to \infty$ in (I.1.27), we get

$$\frac{1}{2}|u_{xx}(\cdot,t)|_2^2 = \frac{1}{2}|u_{xx}(\cdot,0)|_2^2 + R(u).$$

This equality due to Lemmas I.1.9, I.1.10 and embedding theorems immediately implies the continuity of $|u_{xx}(\cdot,t)|_2$ in $t$. Therefore, $\|u(\cdot,t)\|_2$ is a continuous function of $t \in [-T,T]$.

Let us take an arbitrary $t_0 \in [-T,T]$ and a sequence $\{t_n\}_{n=1,2,3,\dots} \subset [-T,T]$ such that $\lim\limits_{n\to\infty} t_n = t_0$. Then, according to the proved facts, the sequence $\{u(\cdot,t_n)\}_{n=1,2,3,\dots}$ converges to $u(\cdot,t_0)$ strongly in $H^1$ and this sequence is weakly compact in $H^2$, therefore any of its limiting points in the space $H^2$ is equal to $u(\cdot,t_0)$ in the weak sense, i. e. this sequence weakly converges to $u(\cdot,t_0)$ in $H^2$. But as it is already proved, $\lim\limits_{n\to\infty} \|u(\cdot,t_n)\|_2 = \|u(\cdot,t_0)\|_2$, hence $\|u(\cdot,t_n) - u(\cdot,t_0)\|_2 \to 0$ as $n \to \infty$. $\square$

**Lemma I.1.16** *For any* $T > 0$   $u_n(\cdot,t) \to u(\cdot,t)$ *as* $n \to \infty$ *in* $C([-T,T]; H^2)$.

**Proof.** Suppose this statement is invalid and there exist $\epsilon > 0$ and a sequence $\{t_n\}_{n=1,2,3,\dots} \subset [-T,T]$ such that

$$\|u_n(\cdot,t_n) - u(\cdot,t_n)\|_2 \geq \epsilon.$$

Let $\{t_{n_k}\}_{k=1,2,3,\dots}$ be a subsequence of the sequence $\{t_n\}_{n=1,2,3,\dots}$ converging to some $t_0 \in [-T,T]$. But $u(\cdot,t_{n_k}) \to u(\cdot,t_0)$ in $H^2$ as $k \to \infty$ and one can easily prove as in Lemma I.1.14 that $u_{n_k}(\cdot,t_{n_k}) \to u(\cdot,t_0)$ as $k \to \infty$ in $H^2$. So, we get a contradiction, and Lemma I.1.16 is proved. $\square$

It easily follows from Lemma I.1.16 that $u_t(\cdot,t) = -f(u(\cdot,t))u_x(\cdot,t) - u_{xxx}(\cdot,t)$ in the sense of the space $H^{-1}$ for any $t \in [-T,T]$ and, thus, we have proved the existence of a generalized solution of the problem (I.1.1),(I.1.2). Let us prove the uniqueness of this solution. Let $u_1(\cdot,t)$ and $u_2(\cdot,t)$ be two generalized solutions of the problem (I.1.1),(I.1.2) defined in an interval of time $(-T_1,T_2)$ where $T_1, T_2 > 0$. We have for $t \in [0,T_2)$ and $w = u_1 - u_2$:

$$\frac{1}{2}\frac{d}{dt}\int\limits_{-\infty}^{\infty} w^2(x,t)dx = -\int\limits_{-\infty}^{\infty} w(f(u_1)u_{1x} - f(u_2)u_{2x})dx =$$

$$= -\int\limits_{-\infty}^{\infty} w[(f(u_1) - f(u_2))u_{1x} + f(u_2)w_x]dx \le C(T_2) \int\limits_{-\infty}^{\infty} w^2 dx$$

and hence $u_1 \equiv u_2$. For $t \in (-T_1, 0]$ the proof can be made by analogy so that the uniqueness of a generalized solution of the problem (I.1.1),(I.1.2) is proved.

To prove Theorem I.1.3, we also need to show the continuous dependence of generalized solutions of the problem (I.1.1),(I.1.2) on the initial data $u_0$. Now, we want to do this.

Let $u_0 \in H^2$ and let a sequence $\{u_0^n\}_{n=1,2,3,\dots} \subset H^2$ converge in this space to $u_0$ as $n \to \infty$. Let $u(x,t)$ and $u_n(x,t)$ be corresponding generalized solutions of the problem (I.1.1),(I.1.2). We need to prove that, for any $T > 0$,

$$\lim_{n \to \infty} \max_{t \in [-T,T]} \|u_n(\cdot,t) - u(\cdot,t)\|_2 = 0. \tag{I.1.29}$$

Let us take an arbitrary $T > 0$. For each number $n$ let $\tilde{u}_0^n \in S$ be such that for the corresponding solution $\tilde{u}_n(x,t) \in C^\infty([-T,T]; S)$ of the problem (I.1.1),(I.1.2) we have

$$\max_{t \in [-T,T]} \|\tilde{u}_n(\cdot,t) - u_n(\cdot,t)\|_2 < \frac{1}{n} \tag{I.1.30}$$

(this point $\tilde{u}_0^n \in S$ exists in view of the above arguments). Then, obviously $\tilde{u}_0^n \to u_0$ in $H^2$ as $n \to \infty$, therefore, as it is proved earlier,

$$\max_{t \in [-T,T]} \|u(\cdot,t) - \tilde{u}_n(\cdot,t)\|_2 \to 0 \quad \text{as } n \to \infty.$$

But then, in view of (I.1.30)

$$\max_{t \in [-T,T]} \|u(\cdot,t) - u_n(\cdot,t)\|_2 \to 0$$

as $n \to \infty$, and the relation (I.1.29) is proved. Theorem I.1.3 is completely proved.□

Now we sketch the proof of Theorem I.1.5. We have $f(u) = u$. Completely as in the case of Proposition I.1.7, one can prove that for any infinitely differentiable $u_0 = u_0(x)$ periodic with the period $A$ there exists a unique solution $u(x,t)$ of the problem (I.1.1),(I.1.2) infinitely differentiable in $x, t \in R$ and periodic in the spatial variable $x$ with the same period $A$ (this statement is also proved in [104]). Below, we use the following result on conservation laws of the problem (I.1.1),(I.1.2).

Lemma I.1.17 *Let $f(u) = u$. Then, there exists a sequence of real functionals defined on $H^\infty = \bigcap\limits_{n=1}^{\infty} H^n_{\text{per}}(A)$ of the kind*

$$E_0(u) = \int\limits_0^A u^2(x)dx, \quad E_1(u) = \int\limits_0^A \left\{ \frac{1}{2} u_x^2 - \frac{1}{6} u^3(x) \right\} dx, \quad E_n(u) =$$

$$= \int_0^A \left\{ \frac{1}{2} \left( \frac{d^n u(x)}{dx^n} \right)^2 + c_n u \left( \frac{d^{n-1} u(x)}{dx^{n-1}} \right)^2 - q_n \left( u(x), ..., \frac{d^{n-2} u(x)}{dx^{n-2}} \right) \right\} dx, \quad , n \geq 2,$$

where $c_n$ are real constants and $q_n$ are polynomials, such that for an arbitrary infinitely differentiable solution $u(x,t)$ of the problem (I.1.1),(I.1.2), periodic in $x$ with the period $A$, the quantities $E_0(u(\cdot,t)), ..., E_n(u(\cdot,t)), ...$ are determined and independent of $t$, i. e. the functionals $E_0, ..., E_n, ...$ are conservation laws for periodic $H^\infty$-solutions of the problem (I.1.1),(I.1.2) with $f(u) = u$.

This statement is related to the complete integrability of the problem (I.1.1),(I.1.2) with $f(u) = u$ and, since the present book is devoted to other questions, we refer readers to the corresponding literature where this result is obtained (see Additional remarks to this chapter).

Let us take arbitrary integer $n \geq 2$, $u_0 \in H^n_{per}(A)$ and a sequence $\{u_0^{(l)}\}_{l=1,2,3,...}$ consisting of infinitely differentiable functions periodic in $x$ with the period $A$ converging to $u_0$ weakly in $H^n_{per}(A)$ as $l \to \infty$. Let $u_l(x,t)$, $l = 1,2,3,...$, be the corresponding infinitely differentiable solutions of the problem (I.1.1),(I.1.2) periodic in $x$ with the period $A$. Then, as in the proof of Theorem I.1.3, one can show that for any $T > 0$ and $t \in (-T,T)$

$$|u(\cdot,t)|_2 = |u(\cdot,0)|_2 \quad \text{and} \quad \max_{t \in [-T,T]} ||u_l(\cdot,t)||_{H^n_{per}(A)} \leq C_1$$

for all $l = 1,2,3,...$ and that there is a limit $u(x,t)$ of the sequence $\{u_l(x,t)\}_{l=1,2,3,...}$ strong in $C((-T,T); H^{n-1}_{per}(A))$ and, for any $t \in R$, weak in $H^n_{per}(A)$; in addition

$$||u(\cdot,t)||_{H^n_{per}(A)} \leq \liminf_{l \to \infty} ||u_l(\cdot,t)||_{H^n_{per}(A)} \leq C_1.$$

Let now $u_0^l \to u_0$ in $H^n_{per}(A)$ strongly and $t_0 \neq 0$ be arbitrary. Since the functionals $E_0, ..., E_n$ are obviously continuous on $H^n_{per}(A)$ and the functionals

$$\int_0^A u^3(x) dx \quad \text{and} \quad \int_0^A \left\{ c_n u \left( \frac{d^{n-1} u}{dx^{n-1}} \right)^2 - q_n \left( u, ..., \frac{d^{n-2} u}{dx^{n-2}} \right) \right\} dx$$

are weakly continuous on this space, we have

$$E_n(u(\cdot,0)) = \liminf_{l \to \infty} E_n(u_l(\cdot,0)) \geq E_n(u(\cdot,t_0));$$

in addition, here the strong equality takes place if and only if $u_l(\cdot,t_0) \to u(\cdot,t_0)$ strongly in $H^n_{per}(A)$ as $l \to \infty$.

Suppose that

$$E_n(u(\cdot,0)) > E_n(u(\cdot,t_0)). \tag{I.1.31}$$

Let $\{u_1^l\}_{l=1,2,3,...}$ be an arbitrary sequence of infinitely differentiable functions of the argument $x \in R$ periodic with the period $A$ and converging to $u(\cdot, t_0)$ as $l \to \infty$ strongly in $H_{\text{per}}^n(A)$. Then, in view of the autonomy of equation (I.1.1), the function $u(\cdot, t)$ is the limit, strong in $C([-T, T]; H_{\text{per}}^{n-1}(A))$ and, for any fixed $t \in [-T, T]$, weak in $H_{\text{per}}^n(A)$ of the corresponding sequence of infinitely differentiable periodic solutions $u_l'(\cdot, t)$, $l = 1, 2, 3, ...$ of equation (I.1.1) satisfying $u_l'(\cdot, t_0) = u_1^l(\cdot)$. Therefore, due to (I.1.31), we have:

$$E_n(u(\cdot, 0)) > E_n(u(\cdot, t_0)) = \liminf_{l \to \infty} E_n(u_l'(\cdot, 0)) \geq E_n(u(\cdot, 0)),$$

i. e. we get a contradiction. Thus, for any $t \in R$ $u_l(\cdot, t) \to u(\cdot, t)$ strongly in $H_{\text{per}}^n(A)$ as $l \to \infty$.

One can prove by the complete analogy that $u(\cdot, t) \in C([-T, T]; H_{\text{per}}^n(A))$ and that, if again $\{u_0^l\}_{l=1,2,3,...} \subset H_{\text{per}}^n(A)$ and $u_0^l \to u_0$ strongly in $H_{\text{per}}^n(A)$ as $l \to \infty$, then $u_l(\cdot, t) \to u(\cdot, t)$ strongly in $C([-T, T]; H_{\text{per}}^n(A))$ for an arbitrary $T > 0$ where $u_l(\cdot, t)$ are corresponding periodic $H^n$-solutions of the problem (I.1.1),(I.1.2).

As in the proof of Theorem I.1.3, $u(\cdot, t)$ is a unique global periodic $H^n$-solution of the problem (I.1.1),(I.1.2) continuously depending on the initial data $u_0$. The statement of Theorem I.1.5 about the time-independence of the quantities $E_0(u(\cdot, t)), ...,$ $E_n(u(\cdot, t))$ follows from the continuity of the functionals $E_0, ..., E_n$ on the space $H_{\text{per}}^n(A)$. Theorem I.1.5 is completely proved.□

## 1.2   The nonlinear Schrödinger equation (NLSE)

In this section, we consider several results on the existence of solutions of the NLSE

$$iu_t + \Delta u + f(|u|^2)u = 0, \quad x \in R^N, \ t \in R \qquad (I.2.1)$$

with prescribed initial data

$$u(x, 0) = u_0(x). \qquad (I.2.2)$$

We shall understand the Laplace operator $\Delta$ in equation (I.2.1) in the generalized sense identifying it with the operator $-D$. Here we shall accept conditions of smoothness for the complex function $f(|u|^2)u : C \longmapsto C$. Considering the complex plane $C$ as the two-dimensional linear space $R^2$, we say that the function $f(|u|^2)u$ is $k$ times continuously differentiable (we write in this case $f(|u|^2)u \in \bigcup_{m,n=1}^{\infty} C^k((-n, n) \times (-m, m); R^2))$) if $f(|u|^2)u$ as a map from $R^2$ into $R^2$ is $k$ times continuously differentiable.

To formulate a result on the existence of solutions of the NLSE vanishing as $|x| \to \infty$, we need the following two assumptions.

(f1) *Let $f(|u|^2)u$ be a continuously differentiable function of the argument $u \in C$ and $f(s)$, where $s \geq 0$, be a real-valued function and let in the case $N \geq 2$ there exist $C_0 > 0$ and $p \in (0, p^* - 1)$, where $p^* = \frac{N+2}{N-2}$ if $N \geq 3$ and $p^* > 1$ is arbitrary for $N = 2$, such that*

$$\left| \frac{\partial}{\partial u}[f(|u|^2)u] \right| \leq C_0(1 + |u|^p), \quad u \in C. \tag{I.2.3}$$

(f2) *Let there exist $C > 0$ and $p_1 \in (0, \frac{4}{N})$ such that $f(s^2) \leq C(1 + s^{p_1})$, $s \in R_+$.*

<u>Remark I.2.1</u> Under $\frac{\partial}{\partial u}[f(|u|^2)u]$ we mean the matrix $\begin{pmatrix} \frac{\partial f(u_1^2+u_2^2)u_1}{\partial u_1} & \frac{\partial f(u_1^2+u_2^2)u_1}{\partial u_2} \\ \frac{\partial f(u_1^2+u_2^2)u_2}{\partial u_1} & \frac{\partial f(u_1^2+u_2^2)u_2}{\partial u_2} \end{pmatrix}$

where $u = u_1 + iu_2$, $u_1, u_2 \in R$.

<u>Definition I.2.2</u> *Let $f(\cdot)$ satisfy condition (f1) and let $T_1, T_2 > 0$ be arbitrary. We call a function $u(\cdot, t) \in C((-T_1, T_2); H^1) \cap C^1((-T_1, T_2); H^{-1})$ a (generalized) solution (or simply a $H^1$-solution) of the problem (I.2.1),(I.2.2) if $u(\cdot, 0) = u_0$ in the sense of the space $H^1$ and equation (I.2.1) after the substitution of this function $u(\cdot, t)$ becomes the equality for any $t \in (-T_1, T_2)$ in the sense of the space $H^{-1}$.*

In this section for the simplicity we present all proofs of the results on the existence for the NLSE for $N = 1$. We convert the problem (I.2.1),(I.2.2) into the following integral equation:

$$u(\cdot, t) = e^{-itD}u_0 + i \int_0^t e^{-i(t-s)D}[f(|u\cdot, s)|^2)u(\cdot, s)]ds. \tag{I.2.4}$$

In fact, the following statement takes place.

<u>Proposition I.2.3</u> *Let $u_0 \in H^1$. Under the assumption (f1) for any $T_1, T_2 > 0$ an arbitrary $H^1$-solution of the problem (I.2.1),(I.2.2) in the interval of time $(-T_1, T_2)$ satisfies equation (I.2.4) in this interval of time. Conversely, a solution $u(\cdot, t) \in C((-T_1, T_2); H^1)$ of equation (I.2.4) is a $H^1$-solution of the problem (I.2.1),(I.2.2) in the interval of time $(-T_1, T_2)$.*

<u>Theorem I.2.4</u> *Under the assumption (f1) for any $u_0 \in H^1$ there exist $T_1, T_2 > 0$ depending only on $\|u_0\|_1$ such that in the interval of time $(-T_1, T_2)$ there exists a unique $H^1$-solution $u(x, t)$ of the problem (I.2.1),(I.2.2) and $T_1 = +\infty$ (resp. $T_2 = +\infty$) if $\limsup_{t \to -T_1+0} \|u(\cdot, t)\|_1 < \infty$ (resp. if $\limsup_{t \to T_2-0} \|u(\cdot, t)\|_1 < \infty$). For any $H^1$-solution the quantities*

$$P(u(\cdot, t)) = \int_{R^N} |u(x, t)|^2 dx, \quad E(u(\cdot, t)) = \int_{R^N} \{\frac{1}{2}|\nabla u(\cdot, t)|^2 - F(|u(\cdot, t)|^2)\}dx$$

*and*

$$M(u(\cdot,t)) = \int_{R^N} u(\cdot,t)\nabla\overline{u}(\cdot,t)dx,$$

where $F(s) = \frac{1}{2}\int_0^s f(r)dr$, are determined and independent of $t$ (i. e. the functionals $P, E, M$ are conservation laws). In addition, under the assumption (f2) any above $H^1$-solution is global in time.

Proof of Proposition I.2.3. Let $I = (-T_1, T_2)$ be an arbitrary interval containing zero and $g(\cdot) \in C(I; H^s)$. Consider the linear problem

$$iw_t - Dw + g(t) = 0, \quad t \in I, \ w(0) = w_0 \in H^s. \tag{I.2.5}$$

We understand its solutions as functions from $C(I; H^s) \cap C^1(I; H^{s-2})$ satisfying the initial condition in $H^s$ and the equation in the sense of $H^{s-2}$ for each $t \in I$. Clearly, the function

$$w(t) = e^{-itD}w_0 + i\int_0^t e^{-i(t-s)D}g(s)ds$$

satisfies the problem (I.2.5). Let us prove the uniqueness of this solution. For this aim, it suffices to show that the homogeneous problem

$$iw_t - Dw = 0, \quad t \in I, \ w(0) = 0 \in H^s \tag{I.2.6}$$

has the unique solution $w(t) \equiv 0$.

Let $w(\cdot) \in C(I; H^s) \cap C^1(I; H^{s-2})$ be a solution of the problem (I.2.6). Consider the function $p(t,r) = e^{-i(t-r)D}w(r) \in C(I; H^s) \cup C^1(I; H^{s-2})$. Then, we have for $r, t \in I, \ 0 \leq r < t$:

$$i\frac{\partial p(t,r)}{\partial r} = i\frac{\partial}{\partial r}\left[e^{-i(t-r)D}\right]w(r) + ie^{-i(t-r)D}\frac{dw(r)}{dr} =$$

$$= -De^{-i(t-r)D}w(r) + e^{-i(t-r)D}Dw(r) = 0$$

in the space $H^{s-2}$, hence $w(t) = p(t,t) = p(t,0) = 0$ and thus $w(r) \equiv 0$, i. e. the uniqueness of the solution of the problem (I.2.5) is proved.

Let $u(\cdot,t) \in C(I; H^1) \cap C^1(I; H^{-1})$ be a solution of the problem (I.2.1),(I.2.2). Then, due to embedding theorems $f(|u(\cdot,t)|^2)u(\cdot,t) \in C(I; H^1)$ (we remind that we consider only the case $N = 1$). Therefore, as in the case of the linear problem (I.2.5) the function $u(\cdot,t)$ satisfies equation (I.2.4). Conversely, let $u(\cdot,t) \in C(I; H^1)$ be a $H^1$-solution of equation (I.2.4). Then, since as earlier $f(|u(\cdot,t)|^2)u(\cdot,t) \in C(I; H^1)$, we have $\frac{\partial}{\partial t}u(\cdot,t) \in C(I; H^{-1})$ and the function $u$ satisfies equation (I.2.1) and the initial condition (I.2.2). Proposition I.2.3 is proved.□

Remark I.1.5 In the proof of Proposition I.2.3 we followed the paper by T. Kato [45] where in fact an essentially more strong result is presented.

Proof of Theorem I.2.4. As noted above, we prove this theorem only for $N = 1$. Also, we assume that $f(|u|^2)u \in \bigcup_{m,n=1}^{\infty} C^2((-m, m) \times (-n, n); R^2)$ for simplicity. Due to Proposition I.2.3, equation (I.2.4) is equivalent to the problem (I.2.1),(I.2.2). So, we consider this equation. We first show the uniqueness of its solution. Let $u_0 \in H^1$ and $u_1(\cdot, t)$ and $u_2(\cdot, t)$ be two solutions of this equation from $C(I; H^1)$. Then, we get by embedding theorems for $t > 0$

$$|u_1(\cdot, t) - u_2(\cdot, t)|_2 \leq C_1 \int_0^t |u_1(\cdot, s) - u_2(\cdot, s)|_2 ds,$$

hence $u_1(\cdot, t) \equiv u_2(\cdot, t)$ for $t > 0$. For $t < 0$ this statement can be proved by analogy, and the uniqueness of a solution of equation (I.2.4) is proved.

Let us prove the existence of a solution. Let $l \geq 1$ be integer. Suppose now that $f(|u|^2)u$ is an infinitely differentiable function. For any $R > 0$ let us show the existence of $T = T(R) > 0$ such that, if $\|u_0\|_l \leq R$, then the operator $G$ in the right-hand side of (I.2.4),

$$(Gu)(t) = e^{-itD}u_0 + i \int_0^t e^{-i(t-s)D}[f(|u\cdot, s)|^2)u(\cdot, s)]ds,$$

maps the set

$$M_T = \{u(\cdot) \in C([-T, T]; H^l) : \ u(0) = u_0 \in H^l, \ \|u(\cdot)\|_l \leq 2\|u_0\|_l\}$$

into itself. Indeed, clearly $G(M_T) \subset C([-T, T]; H^l)$ and, since the operator $\frac{\partial}{\partial x}$ commutes with $e^{-itD}$, by applying embedding theorems we get from (I.2.4) for $u(\cdot, t) \in M_T$:

$$\|(Gu)(t)\|_l \leq \|u_0\|_l + \int_0^t \gamma_l(\|u(\cdot, s)\|_l) ds,$$

where $\gamma_l(s)$ is a continuous positive function of the argument $s \geq 0$. This inequality implies the existence of the above $T > 0$, depending only on $\|u_0\|_l$, for which $G(M_T) \subset M_T$.

Now let us show the existence, for any $R > 0$, of a constant $T_1 \in (0, T]$ such that the map $G$ is a contraction of the set $M_{T_1}$ if $\|u_0\|_l \leq R$. Indeed, using embedding theorems and the infinite differentiability of $f(|u|^2)u$, we easily derive

$$\|(Gu_1)(t) - (Gu_2)(t)\|_l \leq C_2 \int_0^t \|u_1(s) - u_2(s)\|_l ds, \quad u_1(\cdot, t), u_2(\cdot, t) \in M_{T_1},$$

where the constant $C_2 > 0$ depends only on $||u_0||_l$, and therefore such a constant $T_1$ exists.

Since a fixed point of the map $G$ is obviously a solution of equation (I.2.4), we have proved, in the case of the infinite differentiability of the function $f(|u|^2)u$, that for any $u_0 \in H^l$ there exists $T > 0$, depending only on $||u_0||_l$, such that in the interval of time $(-T, T)$ there exists a unique solution $u(\cdot, t) \in C((-T, T); H^l)$ of equation (I.2.4). Since equation (I.2.1) is autonomous (invariant with respect to the substitution $t \to t + c$ where $c$ is a constant), a result similar to the above one on the existence and uniqueness of a $H^l$-solution takes place if we pose the initial condition not at the point $t = 0$ but at an arbitrary point $t = t_0 \neq 0$. Therefore, since the length of the interval of the existence of a $H^l$-solution is bounded from below by a constant $T > 0$ depending only on $||u_0||_l$, we get the existence of $T_l^- < 0$ and $T_l^+ > 0$ such that our $H^l$-solution can be continued onto the interval $(T_l^-, T_l^+)$ and $\limsup_{t \to T_l^- + 0} ||u(\cdot, t)||_l = +\infty$ if $T_l^- > -\infty$ (resp. $\limsup_{t \to T_l^+ - 0} ||u(\cdot, t)||_l = +\infty$ if $T_l^+ < \infty$).

Let again $u_0 \in H^l$, $l \geq 2$. Then obviously for each $k = 1, 2, ..., l$ there exists a unique $H^k$-solution of the problem (I.2.1),(I.2.2) and let $(T_k^-, T_k^+)$, where $T_k^- < 0 < T_k^+$, be the maximal interval of its existence. Then, $-\infty \leq T_1^- \leq T_2^- \leq ... \leq T_l^- < 0$ and $0 < T_l^+ \leq ... \leq T_2^+ \leq T_1^+ \leq +\infty$. Since it follows from equation (I.2.4) that for $k = 1, ..., l - 1$, $T_{k+1} \in (0, T_k^+)$ and $t \in [0, T_{k+1}]$

$$||u(\cdot, t)||_{k+1} \leq ||u_0||_{k+1} + C_3 \left( \max_{t \in [0, T_k]} ||u(\cdot, t)||_k \right) \int_0^t ||u(\cdot, s)||_{k+1} ds$$

and by analogy for $t < 0$, we have $T_1^- = T_2^- = ... = T_l^-$ and $T_1^+ = T_2^+ = ... = T_l^+$.

Let $l \geq 3$, $u_0 \in H^1$, $\{u_0^n\}_{n=1,2,3,...} \subset H^l$, $u_0^n \to u_0$ as $n \to \infty$ in $H^1$, and let $\{f^n(|u|^2)u\}_{n=1,2,3,...}$ be a sequence of infinitely differentiable functions satisfying condition (I.2.3) with the same constants $C$ and $p$ and such that for any $k, m = 1, 2, 3, ...$ $f^n(|u|^2)u \to f(|u|^2)u$ as $n \to \infty$ in $C^2((-k, k) \times (-m, m); R^2)$. Let $u(\cdot, t)$ be a $H^1$-solution and $u_n(\cdot, t)$ $(n = 1, 2, 3, ...)$ be the sequence of $H^l$-solutions of the problem (I.2.1),(I.2.2) taken with $u_0 = u_0^n$ and $f(\cdot) = f^n(\cdot)$. Let $(-T_1, T_2)$ and $(-T_1^n, T_2^n)$ be maximal intervals of the existence of these solutions, respectively.

Let us take arbitrary $t' \in (0, T_1)$ and $t'' \in (0, T_2)$. We want to prove that $T_1^n > t'$ and $T_2^n > t''$ for all sufficiently large numbers $n$ and that $u_n(\cdot, t) \to u(\cdot, t)$ as $n \to \infty$ in $C([-t', t'']; H^1)$. In view of the above arguments there exist $t_1 \in (0, t']$ and $t_2 \in (0, t'']$ such that for all sufficiently large numbers $n$ $H^l$-solutions $u_n(\cdot, t)$ can be continued onto the segment $[-t_1, t_2]$ and are bounded in $H^l$ uniformly with respect to $t \in [-t_1, t_2]$. Then, for $t \in [-t_1, t_2]$ we have $(t > 0)$:

$$|u(\cdot, t) - u_n(\cdot, t)|_2 \leq C_n^1 + C_4 \int_0^t |u(\cdot, s) - u_n(\cdot, s)|_2 ds,$$

where $C_n^1 \to +0$ as $n \to \infty$ and $C_4 = \text{constant} > 0$ and by analogy for $t < 0$, hence $u_n(\cdot, t) \to u(\cdot, t)$ as $n \to \infty$ in $C([-t_1, t_2]; L_2)$.

Further, by embedding theorems

$$\left| \frac{\partial}{\partial x}(u(\cdot, t) - u_n(\cdot, t)) \right|_2 \leq C_n^2 + C_5 \int_0^t \left| \frac{\partial}{\partial x}(u(\cdot, s) - u_n(\cdot, s)) \right|_2 ds,$$

where $C_n^2 \to +0$ as $n \to \infty$ and $C_5 = \text{constant} > 0$ and by analogy for $t < 0$, therefore $u_n(\cdot, t) \to u(\cdot, t)$ as $n \to \infty$ in $C([-t_1, t_2]; H^1)$. The latter fact implies, in particular, the convergence of the sequences $\{\|u_n(\cdot, t_i)\|_1\}_{n=1,2,3,\dots}$, $i = 1, 2$, to $\|u(\cdot, t_i)\|_1$, respectively. Hence, if $t_1 < t'$ or $t_2 < t''$, then there exists $t_1' \in (t_1, t']$ (resp. $t_2' \in (t_2, t'']$) depending only on $\|u(\cdot, t_1)\|_1$ (resp. on $\|u(\cdot, t_2)\|_1$) such that for all sufficiently large numbers $n$ solutions $u_n(\cdot, t)$ can be continued onto the segment $[-t_1', 0]$ (resp., on $[0, t_2']$) and, as above, converge to $u(\cdot, t)$ as $n \to \infty$ in $C([-t_1', 0]; H^1)$ (resp., in $C([0, t_2']; H^1)$). Continuing this process, since $\max_{t \in [-t', t'']} \|u(\cdot, t)\|_1 < \infty$, after a finite number of steps we shall get that $u_n(\cdot, t) \to u(\cdot, t)$ as $n \to \infty$ in $C([-t', t'']; H^1)$.

The continuous dependence of a $H^1$-solution of the problem (I.2.1),(I.2.2) on its initial data can be proved on the complete analogy of this construction.

Since $P, E, M$ are obviously continuously differentiable functionals on the space $H^1$ and since the direct verification shows the independence of $t$ of the quantities $P(u(\cdot, t)), E(u(\cdot, t))$ and $M(u(\cdot, t))$ for an arbitrary $H^l$-solution $u(\cdot, t)$ of the problem (I.2.1),(I.2.2) $(l \geq 3)$, according to the above arguments these quantities are also independent of $t$ if $u(\cdot, t)$ is a $H^1$-solution of this problem.

Finally, under the assumption (f2) using inequality (I.1.15), we get

$$\int_{-\infty}^{\infty} F(|u(x)|^2) dx \leq C_6 |u|_2^2 + C_7 |u|_2^{\frac{p+2}{2}+1} |u_x|_2^{\frac{p+2}{2}-1},$$

where $0 < \frac{p+2}{2} - 1 < 2$, hence, if $u_0 \in H^1$, then, since $E(u(\cdot, t)) = E(u(\cdot, 0))$ and $|u(\cdot, t)|_2 = |u(\cdot, 0)|_2$ for any $t$, there exists $C' > 0$ such that $|u_x(\cdot, t)|_2 \leq C'$ for all $t \in R$ from the interval of the existence of the corresponding solution $u(\cdot, t)$ of the problem (I.2.1),(I.2.2). Therefore, this solution is global, and Theorem I.2.4 is proved. □

Remark I.2.6 With the proof of Theorem I.2.4 we in fact have shown for a sufficiently smooth function $f$ the existence of $H^l$-solutions (where $l = 2, 3, 4, \dots$) of equation (I.2.4) if $u_0 \in H^l$.

Now we consider a result on the existence of solutions for the NLSE in classes of functions nonvanishing as $|x| \to \infty$. First of all, it should be noted that the Cauchy problem for the linear one-dimensional NLSE with $f \equiv 0$ and $N = 1$ is ill-posed in the space $C(R)$. Indeed, this follows from the fact that the function $\sqrt{\frac{t_0}{t_0-t}}e^{-\frac{ix^2}{4(t_0-t)}}$ with an arbitrary $t_0 > 0$ for each $t \in [0, t_0)$ belongs to $C(R)$ as a function of the argument $x \in R$ and satisfies our linear NLSE because this function is unbounded in any left half-neighborhood of the point $t = t_0$ (this example is taken from [80]). We shall show the well-posedness of the problem (I.2.1),(I.2.2) in the spaces $X^k$ (i. e. when for initial data $u_0$ additional conditions of the regularity are assumed).

It is known that the operator $e^{-itD}$ acts in the spaces $L_2$ and $H^k$, $k = 1, 2, 3, ...,$ as the following integral operator:

$$G_t\phi = \int\limits_{-\infty}^{\infty} K(x-y,t)\phi(y)dy \ (t \neq 0), \ G_0\phi \equiv \phi,$$

where for $t \neq 0$   $K(x,t) = (4\pi it)^{-\frac{1}{2}} \exp(\frac{ix^2}{4t})$ (here the root $(4\pi it)^{\frac{1}{2}}$ lies in the first quadrant Re$z > 0$, Im$z > 0$ if $t > 0$ and in the fourth quadrant Re$z > 0$, Im$z < 0$ if $t < 0$). Of course, the function $K(x,t)$ is the fundamental solution of the operator $L = i\frac{\partial}{\partial t} + \frac{\partial^2}{\partial x^2}$. We also set $g(u) = f(|u|^2)u$. In this way the integral equation (I.2.4) formally can be written as follows:

$$u(x,t) = G_t u_0 + i\int\limits_{0}^{t} G_{t-s}g(u(\cdot,s))ds. \tag{I.2.7}$$

We begin with the following statement.

**Proposition I.2.7** *For any $k = 1, 2, 3, ...$ the operators $G_t$ considered on $X^k$ satisfy the following properties:*

*(G1) for any bounded interval $I \subset R$ the family of operators $G_t :  X^k \to X^k$ is uniformly bounded with respect to $t \in I$;*

*(G2) for any $\phi \in X^k$ the function $G_t\phi :  I \to X^k$ is continuous and $\lim\limits_{t\to 0} G_t\phi = \phi$ in the sense of the space $X^k$.*

Proof. Let $t \neq 0$. Consider $\phi \in X^k$ and

$$G_t\phi = (\pi i)^{-\frac{1}{2}}\{\int\limits_{-\infty}^{-\beta} e^{iz^2}\phi(x+2\sqrt{t}z)dz + \int\limits_{-\beta}^{\beta} e^{iz^2}\phi(x+2\sqrt{t}z)dz + \int\limits_{\beta}^{\infty} e^{iz^2}\phi(x+2\sqrt{t}z)dz\} =$$

$$= (\pi i)^{-\frac{1}{2}}\{I_1 + I_2 + I_3\}, \tag{I.2.8}$$

where $\beta > 0$ is arbitrary (in what follows, we prove that the improper integrals here are converging). We have

$$|I_1| \leq \left| \lim_{\alpha \to +\infty} \left[ ie^{is} \frac{\phi(x - 2\sqrt{ts})}{2\sqrt{s}} \right]_{s=\beta^2}^{s=\alpha^2} \right| +$$

$$+\frac{1}{2} \lim_{\alpha \to +\infty} \left| \int_{\beta^2}^{\alpha^2} e^{is} \left\{ \frac{\sqrt{t}\phi'(x - 2\sqrt{ts})}{s} + \frac{\phi(x - 2\sqrt{ts})}{2s^{\frac{3}{2}}} \right\} ds \right| \leq$$

$$\leq C_1 \beta^{-1} |\phi|_C + \sqrt{t} \lim_{\alpha \to +\infty} \int_{-2\alpha\sqrt{t}}^{-2\beta\sqrt{t}} \frac{|\phi'(x + z)|}{|z|} dz \leq$$

$$\leq C_1 \beta^{-1} |\phi|_C + C_2 \beta^{-\frac{1}{2}} t^{\frac{1}{4}} |\phi'|_2. \tag{I.2.9}$$

By analogy,

$$|I_3| \leq C_3 \beta^{-1} |\phi|_C + C_4 \beta^{-\frac{1}{2}} t^{\frac{1}{4}} |\phi'|_2. \tag{I.2.10}$$

Thus, by (I.2.8)-(I.2.10)

$$|G_t \phi|_C \leq C_5 |||\phi|||_k$$

for all $t \in I$.

Further, since for $\phi \in C_0^\infty(R)$, we get

$$\frac{d}{dx}[G_t \phi] = G_t \phi'$$

and, since the operator $G_t : L_2 \to L_2$ is unitary, for any $\phi \in X^k$ and a finite interval $I \in R$ we have

$$\left| \frac{d^l}{dx^l} G_t \phi \right|_2 \leq |||\phi|||_k \quad (l = 1, 2, ..., k),$$

and the statement (G1) follows.

Concerning (G2), we only prove that $G_t \phi \to \phi$ as $t \to 0$ in $X^k$ because the other part of this statement can be proved by analogy. Fix an arbitrary $\epsilon > 0$. According to (I.2.9),(I.2.10), there exists $\beta > 0$ such that for all $t : |t| < 1$ the following two inequalities take place:

$$\pi^{-\frac{1}{2}}(|I_1| + |I_3|) < \frac{\epsilon}{3(k+1)} \quad \text{and} \quad \left| (i\pi)^{-\frac{1}{2}} \int_{-\beta}^{\beta} e^{ix^2} dx - 1 \right| < \frac{\epsilon}{3(k+1)} \tag{I.2.11}$$

(the last one is valid because $(i\pi)^{-\frac{1}{2}} \int_{-\infty}^{\infty} e^{ix^2} dx = 1$).

Then, the function $\phi(x + 2\sqrt{t}z)$ converges to $\phi(x)$ as $t \to 0$ uniformly with respect to $z \in [-\beta, \beta]$. Therefore, there exists $t_0 > 0$ such that if $|t| < t_0$, then

$$|(i\pi)^{-\frac{1}{2}} \int_{-\beta}^{\beta} e^{iz^2} \phi(x + 2\sqrt{t}z)dz - \phi(x)| < \frac{\epsilon}{3(k+1)}. \tag{I.2.12}$$

Further, by the strong continuity of the unitary group $e^{itD}$ in the space $L_2$

$$|\frac{d^l}{dx^l} G_t\phi - \phi^{(l)}|_2 < \frac{\epsilon}{k+1} \quad (l = 1, 2, ..., k) \tag{I.2.13}$$

for sufficiently small $t$. By (I.2.11)-(I.2.13) $|||G_t\phi - \phi|||_k < \epsilon$ for all sufficiently small $t$ and Proposition I.2.7 is proved. $\square$

**Proposition I.2.8** *Let $N = 1$, $g(u)$ be a complex-valued infinitely differentiable function of the complex argument $u$ and $I$ be an interval containing zero. Then, a function $u(\cdot, t) \in C(I; X^3) \cap C^1(I; X^1)$ satisfies the problem*

$$i\frac{\partial}{\partial t}u + \frac{\partial^2}{\partial x^2}u + g(u) = 0, \quad x \in R, \ t \in I, \tag{I.2.14}$$

$$u(x, 0) = u_0 \in X^3 \tag{I.2.15}$$

*if and only if $u(x, 0) = u_0(x)$ for any $x \in R$ and this function is a solution of equation (I.2.7).*

    <u>Proof.</u> Let first a function $u(\cdot, t) \in C(I; X^3)$ satisfy equation (I.2.7). We shall show that it is a solution of the problem (I.2.14),(I.2.15). For this aim, we have to prove that this function satisfies equation (I.2.14).

    Clearly, $g(u(\cdot, t)) \in C(I; X^3)$. Let us show that $L[G_t u_0] = 0$ for $t \neq 0$. We have

$$\frac{\partial^2}{\partial x^2}[G_t u_0] = \frac{\partial^2}{\partial x^2} \int_{-\infty}^{\infty} K(y, t)u_0(x + y)dy = G_t\left(\frac{d^2 u_0}{dx^2}\right),$$

where the right-hand side is continuous.

    Let us show the existence of $\frac{\partial}{\partial t}[G_t u_0]$. By setting $z = \frac{(x-y)^2}{4t}$ and accepting for the definiteness that $t > 0$ (the case $t < 0$ can be considered by analogy), we get

$$G_t u_0 = (4\pi i)^{-\frac{1}{2}} \int_0^{\infty} e^{iz} \frac{u_0(x + 2\sqrt{t}z) + u_0(x - 2\sqrt{t}z)}{\sqrt{z}}dz.$$

This relation yields formally

$$\frac{\partial}{\partial t}[G_t u_0] = (4\pi i)^{-\frac{1}{2}} \int_0^{\infty} e^{iz} \frac{u_0'(x + 2\sqrt{t}z) - u_0'(x - 2\sqrt{t}z)}{\sqrt{t}}dz.$$

For $c > 0$ we have

$$\int\limits_0^c e^{iz} \frac{u_0'(x + 2\sqrt{tz}) - u_0'(x - 2\sqrt{tz})}{\sqrt{t}} dz = -ie^{ic}\frac{u_0'(x + 2\sqrt{tc}) - u_0'(x - 2\sqrt{tc})}{\sqrt{t}} +$$

$$+i\int\limits_0^c e^{iz} \frac{u_0''(x + 2\sqrt{tz}) + u_0''(x - 2\sqrt{tz})}{\sqrt{z}} dz,$$

where $u_0'(x) \to 0$ as $|x| \to \infty$ and, consequently, the first term in the right-hand side of this equality tends to zero as $c \to \infty$ uniformly with respect to $t$ from any bounded interval. Since due to estimates (I.2.8)-(I.2.10) the second term in the right-hand side is an improper integral which converges uniformly in $t$ from any bounded interval not containing zero, for any $t > 0$ and $x \in R$ the derivative $\frac{\partial}{\partial t}[G_t u_0]$ is determined and

$$\frac{\partial}{\partial t}[G_t u_0] = i(4\pi i)^{-\frac{1}{2}} \int\limits_0^\infty e^{iz} \frac{u_0''(x + 2\sqrt{tz}) + u_0''(x - 2\sqrt{tz})}{\sqrt{z}} dz =$$

$$= i(4\pi it)^{-\frac{1}{2}} \int\limits_{-\infty}^\infty e^{i\frac{(x-y)^2}{4t}} u_0''(y) dy = i\frac{\partial^2}{\partial x^2}[G_t u_0]$$

so that indeed $L[G_t u_0] \equiv 0$ for $t > 0$. As we noted earlier, the proof of this relation for $t < 0$ can be made by analogy.

If $t = 0$, then obviously $\frac{\partial^2}{\partial x^2}[G_t u_0] = u_0''$. Further, due to the above arguments $\frac{\partial}{\partial t}[G_t u_0] = iG_t u_0''$ and therefore there exists $\lim\limits_{t\to 0} \frac{\partial}{\partial t}[G_t u_0] = iu_0''$. Hence, there exists $\frac{\partial}{\partial t}[G_t u_0]\Big|_{t=0} = iu_0''$ and $L[G_t u_0]\Big|_{t=0} = 0$. These arguments also imply that

$$L\left\{i\int\limits_0^t G_{t-s}(g(u(\cdot, s)))ds\right\} = -g(u(x, t)).$$

Thus, our first statement that any $X^3$-solution of equation (I.2.7) satisfies the problem (I.2.14),(I.2.15) is proved.

Now we prove that any $X^3$-solution of the problem (I.2.14),(I.2.15) satisfies equation (I.2.7). In view of the above arguments, it suffices to prove that the linear homogeneous problem

$$Lu = 0, \quad x \in R, \ t \in I,$$

$$u(x, 0) \equiv 0$$

has only the trivial solution $u \equiv 0$ from $C(I; X^3)$. Let us suppose that $u(x, t)$ is a $X^3$-solution of this problem in the interval of time $I$. Simple calculations show that

$$\frac{d}{dt}\int\limits_{-\infty}^\infty |u_x(x, t)|^2 dx = 0$$

for all $t \in I$. Therefore, the function $u(x,t)$ is constant in $x$ for any fixed $t$. But then, in view of the equation, $u(x,t) \equiv 0$. Thus, Proposition I.2.6 is proved.$\Box$

Now we accept the following definition natural in view of Proposition I.2.8.

<u>Definition I.2.9</u> *Let $N = 1$, $k = 1$ or $k = 2$, $g(u)$ be a complex-valued continuous function of the complex argument $u$ and $I$ be an interval containing zero. We call a solution $u(\cdot,t) \in C(I; X^k)$ of equation (I.2.7) a generalized solution (or a $X^k$-solution) of the problem (I.2.14),(I.2.15).*

Our result on the existence for the NLSE in classes of functions nonvanishing as $|x| \to \infty$ is the following.

<u>Theorem I.2.10</u> *Let $k \geq 1$ be integer and $g(\cdot)$ be a complex-valued $(k+1)$ times continuously differentiable function of the complex argument. Then, for any $u_0 \in X^k$ there exist $T > 0$, depending only on $|||u_0|||_k$, and a unique $X^k$-solution $u(\cdot,t)$ of the problem (I.2.14),(I.2.15) defined in the interval of time $(-T,T)$. Further, there exist $T_1^*, T_2^* > 0$ such that this solution can be continued on the interval $(-T_1^*, T_2^*)$ and $T_1^* = \infty$ (resp. $T_2^* = \infty$) if $\limsup_{t \to -T_1^*+0} |||u(\cdot,t)|||_k < +\infty$ (resp. if $\limsup_{t \to T_2^*-0} |||u(\cdot,t)|||_k < +\infty$). This solution continuously depends on $u_0 \in X^k$ in the sense that for any $T_1 \in (0, T_1^*)$, $T_2 \in (0, T_2^*)$ and a sequence $\{u_0^{(n)}\}_{n=1,2,3,\ldots} \subset X^k$ the sequence of corresponding $X^k$-solutions $u_n(\cdot,t)$ of the problem (I.2.14),(I.2.15) taken with $u_0 = u_0^{(n)}$ converges to $u(\cdot,t)$ in $C([-T_1, T_2]; X^k)$ as $n \to \infty$.*

<u>Proof</u> here repeats the proof of the Theorem I.2.4.$\Box$

## 1.3   On the blowing up of solutions

The phenomenon indicated in the title is known for the NLSE and unknown for the KdVE. It consists in the property of a solution to be unbounded (in a functional space, of course) in a bounded interval of time. It was discovered long time ago. Evidently, one of the first mathematical results in this direction for the NLSE was presented in the paper [38]. Here, we only touch upon this subject. We restrict our attention to the case $f(|u|^2) = |u|^p$. Also, we do not present complete proofs but carry out only a formal technique.

Let $y = \int_{R^N} |x|^2 |u|^2 dx$ and $z = \text{Im} \int_{R^N} r u \bar{u}_r dx$ where $r = |x|$. We consider $H^1$-solutions of the NLSE. If $u$ is a solution of the NLSE (I.2.1), then one formally has:

$$\frac{d}{dt}y = -4z, \quad \frac{d}{dt}z = -2|\nabla u|_2^2 + \frac{pN}{p+2}\int\limits_{R^N}|u|^{p+2}dx. \tag{I.2.16}$$

Identities (I.2.16) imply:

$$\frac{d}{dt}z = \frac{pN-4}{2}|\nabla u|_2^2 - pNE(u_0). \tag{I.2.17}$$

Furthermore, the inequality

$$|z| \leq \left\{\int\limits_{R^N}|x|^2|u|^2dx\right\}^{\frac{1}{2}}|\nabla u|_2 \tag{I.2.18}$$

is valid. If $E(u_0) < 0$, $p > \frac{4}{N}$ and $z(0) \geq 0$, then by (I.2.16) and (I.2.17) $y$ is a decreasing positive function, hence we obtain from (I.2.17),(I.2.18):

$$\frac{d}{dt}z \geq \frac{pN-4}{2}[y(0)]^{-1}z^2 - pNE(u_0). \tag{I.2.19}$$

Thus, by simple arguments there exists $T > 0$ such that $z \to +\infty$ as $t \to T - 0$. By (I.2.17) we also have $\lim\limits_{t \to T-0}|\nabla u|_2 = +\infty$.

So, we have shown the blow up of a solution $u(x,t)$ of the NLSE with $f(|u|^2) = |u|^p$ if $E(u(0)) < 0$, $z(0) \geq 0$ and $p > \frac{4}{N}$.

## 1.4 Additional remarks

The results on the well-posedness of the Cauchy problem for the KdVE and NLSE considered in this chapter are far from being complete. Our approach to proving Theorem I.1.3 is close to the method from [29] (see also [51]). In [44] T. Kato, using the method of semigroups of operators, besides other results, has proved the well-posedness for $u_0 \in H^s$, local for $s > \frac{3}{2}$ and global for $s \geq 2$, of the Cauchy problem for the KdVE under suitable assumptions on $f$. In [46] for the standard KdVE with $f(u) = u$ the well-posedness in $H^s$, local for $s > \frac{3}{4}$ and global for $s \geq 1$, is proved. In [29,44,51] some smoothness properties of solutions are also investigated. In the papers [64] and [49], the Cauchy problem for the (integrable) KdVE with $f(u) = u$ is studied when initial data $u_0$ are nonvanishing as $|x| \to \infty$. The problem of the existence of an infinite sequence of conservation laws for the integrable KdVE (see Theorem I.1.5) is investigated in many papers (see, for example, [50,94]); in [50] it is proved that these quantities have the form indicated in the formulation of this theorem. In addition, an interesting result is obtained by A. Cohen and T. Kappeler in [26] who proved that the KdVE with $f(u) = u$ and with zero initial data $u_0(x) \equiv 0$ has a $C^\infty$- solution which is not identically equal to zero.

Concerning the NLSE, the proof of Theorem I.2.4 in its general form is contained in the paper [45]. For this equation, there are a lot of investigations of the well-posedness of the Cauchy problem with initial data $u_0$ from $H^1$ or more smooth (see, for example, [33,37,69,70,79,88]). With Theorem I.2.4 we have proved one of the simplest results on the existence of $H^1$-solutions (we recall that we considered only the one-dimensional case). Proposition I.2.7 and Theorem I.2.10 are first proved in [102]. We also especially mention the important result of Y. Tsutsumi [91] stating for $f(s) = \pm s^p$ with some $p \in (0, \frac{2}{N})$ the well-posedness of the Cauchy problem for the NLSE (I.2.1),(I.2.2) with initial data from $L_2$.

For the equations under consideration with non-smooth initial data, there is an essential difference between investigations of the well-posedness of the problem periodic in $x$ (with $N = 1$ for the NLSE) and the problem with initial data vanishing as $|x| \to \infty$. Having no possibility to review the whole literature devoted to the periodic problem, we mention the paper by J. Bourgain [16] (see also [17]) where the existence and uniqueness of $L_2$-solutions periodic in $x$ are proved for the usual (with $f(u) = u$) KdVE and for the NLSE with superlinear nonlinearities like $f(s) = |s|^p$.

As for the phenomenon of the blow up of solutions, it is known for the NLSE but is unknown for the KdVE. Above we have considered one of the simplest results on this subject and followed the paper [38]. Although we have presented only a formal technique as in [38], this result is justified rigorously (see, for example, [90] and, also, [69,70]). For $f(s) = s^p$ with $p = \frac{2}{N}$, it is proved in the paper [90] that there exist blowing up solutions, too.

# Chapter 2

# Stationary problems

As we have already noted in the Introduction, the substitution of the general representation for standing waves $u(x,t) = e^{i\omega t}\phi(x)$, $\omega \in R$, into the NLSE leads to the stationary equation

$$\Delta\phi - \omega\phi + f(\phi^2)\phi = 0, \quad x \in R^N, \quad \phi = \phi(x). \tag{II.0.1}$$

Here $\phi$ is a real-valued function. We also consider a generalization of this equation,

$$\Delta\phi - \omega\phi + f(x, \phi^2)\phi = 0, \quad x \in R^N, \tag{II.0.2}$$

and supply equations (II.0.1) and (II.0.2) with some conditions on the infinity for their solutions. Usually we suppose that solutions $\phi(x)$ vanish as $|x| \to \infty$, i. e.

$$\phi(x)\big|_{|x|\to\infty} = 0. \tag{II.0.3}$$

The problem of finding solitary waves for the KdVE under natural assumptions on the behavior of these solutions as $|x| \to \infty$ can be reduced to equation (II.0.1) with $N = 1$, too. Along with the problems (II.0.1),(II.0.3) and (II.0.2),(II.0.3) we also consider the following similar problem:

$$\Delta\phi - \omega\phi + f(x, \phi^2)\phi = 0, \quad x \in \Omega, \quad \phi = \phi(x), \tag{II.0.4}$$

$$\phi\big|_{\partial\Omega} = 0, \tag{II.0.5}$$

where $\Omega \subset R^N$ is now a bounded domain with a sufficiently smooth boundary.

Suppose that $N \geq 3$ and $f(x, \phi^2) = k(x)|\phi|^{p-1}$, $p > 1$, where $k(x)$ is a $C^1$-function and $k(x) > 0$ in $\overline{\Omega}$. It is known that if $p \in \left(1, \frac{N+2}{N-2}\right)$ and $\omega \geq 0$, then the problem (II.0.4),(II.0.5) has a solution positive in $\Omega$ and an infinite sequence of pairwise different solutions. If $f = |\phi|^{p-1}$, $p \geq \frac{N+2}{N-2}$, $\omega \geq 0$ and the domain $\Omega$ is starshaped, then, as it is proved in [74], this problem has no nontrivial solutions. By analogy, as it will be shown further (see Example II.0.1 and Section 2 of this chapter), the problem (II.0.1),(II.0.3) with $N \geq 3$, $f(\phi^2) = \phi^{p-1}$, $p > 1$, and $\omega > 0$ has nontrivial solutions if and only if $p < \frac{N+2}{N-2}$. The exponent $p^* = \frac{N+2}{N-2}$ is called critical in the literature. Respectively, the nonlinearity $f(\phi^2)\phi$ (or $f(x, \phi^2)\phi$) satisfying the condition $\lim\limits_{|\phi| \to \infty} f(\phi^2) = +\infty$ is called subcritical if $\lim\limits_{|\phi| \to \infty} \frac{f(\phi^2)}{|\phi|^{\frac{4}{N-2}}} = 0$ and supercritical if $\lim\limits_{|\phi| \to \infty} \frac{f(\phi^2)}{|\phi|^{\frac{4}{N-2}}} = +\infty$. Roughly speaking, the problem (II.0.4),(II.0.5) (or (II.0.1),(II.0.3)) with $\omega > 0$ and a subcritical function $f(\cdot)$ has a positive solution and an infinite sequence of pairwise different (resp. radial i. e. depending only on $r = |x|$) solutions (on this subject see, for example, [1,77,78] and Theorem II.2.1). Problems with superlinear nonlinearities $f(\cdot)$ have been intensively handled, too, and there have been obtained many interesting and important results in this direction. However, in the present book we shall not deal with supercritical nonlinearities.

In the literature, there are a lot of investigations of the existence of radial solutions of the problem (II.0.1),(II.0.3) that we shall deal with. As for the non-radial problem (II.0.2),(II.0.3), we mention here the papers [57,58] among publications devoted to it. We illustrate some specific features of this problem with the following simple example taken from [99].

Example II.0.1 Let $f(x, \phi^2) = k(x_1)|\phi|^{p-1}$, $N \geq 3$, $\omega > 0$ and $p > 1$, where $k(\cdot)$ is a $C^1$-function, $k'(x_1) > 0$ for all $x_1 \in R$, the derivative $k'(x_1)$ sufficiently rapidly tends to zero as $|x_1| \to \infty$ and $0 < C_1 \leq k(x_1) < C_2 < +\infty$ for some positive constants $C_1$ and $C_2$ independent of $x_1 \in R$. We recall that, as it is already mentioned above, under similar assumptions the problem (II.0.4),(II.0.5) in a bounded domain with $p < \frac{N+2}{N-2}$ has nontrivial solutions. Here we show that the problem (II.0.2),(II.0.3) has no nontrivial solutions. Let $\phi(x)$ be a solution of this problem. Using, for example, the maximum principle, one can easily get the estimate

$$|\phi(x)| + \sum_{k=1}^{N} \left| \frac{\partial \phi(x)}{\partial x_k} \right| + |\Delta\phi(x)| \leq C_3 e^{-C_4|x|} \tag{II.0.6}$$

with some positive constants $C_3$ and $C_4$ independent of $x \in R^N$. Multiplying equation (II.0.2) by $\frac{\partial \phi(x)}{\partial x_1}$ and integrating the obtained equality over $R^N$ with the use of integration by parts, we get

$$\int_{R^N} k'(x_1)|\phi(x)|^{p+1}dx = 0.$$

Hence, since $k'(x_1) > 0$, we have $\phi(x) \equiv 0$.

So, we shall usually look for radial solutions of the problem (II.0.1),(II.0.3). In this connection, the important results by B. Gidas, Ni Wei-Ming and L. Nirenberg

[34,35] should be mentioned: these authors have proved that an arbitrary positive solution of the problem (II.0.4),(II.0.5) in a ball with a function f independent of $x$ and of the problem (II.0.1),(II.0.3) under assumptions of general type is radial. As it is shown in the above papers, for solutions with alternating sign this is not so. Now we also present (without a proof) the result by H. Berestycki and P.L. Lions from [9] on the existence of radial solutions of a problem similar to (II.0.1),(II.0.3) under assumptions of a very general kind.

<u>Theorem II.0.2</u> *Let* $g(\cdot) \in \bigcup_{n=1}^{\infty} C((-n,n);R)$ *be odd,* $N \geq 3$ *and let*

*(a)* $0 < \liminf\limits_{u \to 0} \frac{g(u)}{u} \leq \limsup\limits_{u \to 0} \frac{g(u)}{u} < +\infty;$

*(b)* $\liminf\limits_{|u| \to +\infty} \frac{g(u)}{u^{p^\star}} \geq 0$ *where again* $p^\star = \frac{N+2}{N-2};$

*(c) there exist* $\eta \in (0, +\infty)$ *such that* $G(\eta) > 0$, *where* $G(\phi) = -2 \int_0^\phi g(s)ds.$

*Then, the problem*

$$\Delta\phi = g(u), \quad u = u(|x|), \quad x \in R^N, \quad \phi\big|_{|x| \to \infty} = 0$$

*has a countable set of pairwise different radial solutions and a positive radial solution.* In this chapter, we shall prove this Theorem in several particular cases and study the qualitative behavior of solutions.

Now, we also establish the <u>Pohozaev identity</u> for the problem (II.0.1),(II.0.3) taken in the same form as in [87]:

$$(N-2) \int_{R^N} |\nabla\phi|^2 dx = -(N-2) \int_{R^N} \phi g(\phi)dx = N \int_{R^N} G(\phi)dx$$

where $g(\phi) = \omega\phi - f(\phi^2)\phi$ and $G(\phi) = -2\int_0^\phi g(s)ds$. A similar equality was also obtained in [74] for the problem (II.0.4),(II.0.5) in the case when the domain $\Omega$ is bounded and the function $f$ is independent of $x$. Of course, in our case of the problem (II.0.1),(II.0.3) one needs additional assumptions on the function $f(\phi^2)$ in the equation. For example, $\omega > 0$ and $f(\phi^2) = |\phi|^{p-1}$, $p > 1$, are sufficient. To get the above equality in this case, taking into account (II.0.6) one may, first, multiply equation (II.0.1) by $\phi$ with the further integration of the result over $R^N$ and, second, multiply the same equation by $x_i \frac{\partial\phi(x)}{\partial x_i}$, sum over $i = 1, 2, ..., N$ and integrate the result over $R^N$ with the use of integration by parts.

The Pohozaev identity in particular yields that the problem (II.0.1),(II.0.3) has only the trivial solution $\phi(x) \equiv 0$ if the function $NG(\phi) + (N-2)\phi g(\phi)$ does not change the sign. This is valid, for example, for $\omega > 0$ and $f(\phi^2) = |\phi|^{p-1}$ with $p \geq \frac{N+2}{N-2}$ $(N \geq 3)$.

## 2.1  Existence of solutions. An ODE approach

At first, we consider solitary waves for the KdVE and for the NLSE with $N = 1$. The substitution $u(x,t) = \phi(x - ct)$, $c \in R$, into the KdVE leads to the equation

$$-\omega\phi' + f(\phi)\phi' + \phi''' = 0.$$

Assuming that $\phi(\pm\infty) = a_\pm$, where $a_+$ and $a_-$ are constants, and that $\phi''(\pm\infty) = 0$, we come to the following equation:

$$-\omega\phi + \overline{f}(\phi) + \phi'' = a \qquad\qquad (\text{II.1.1})$$

with $a = -\omega a_-$ and $\overline{f}(\phi) = \int_{a_-}^{\phi} f(s)ds$. It is clear that the substitution of the general representation for solitary waves into the NLSE leads to a similar equation. Of course, equation (II.1.1) can be solved by quadratures. However, it is simpler to make a qualitative analysis we consider here.

First of all, we note that if $\overline{f}$ is a continuously differentiable function and a solution of this equation is bounded on a half-interval $[a, b)$, where $a < b$, then it can be continued onto a right half-neighborhood of the point $b$. Indeed, if a solution $\phi$ of equation (II.1.1) is bounded, then it follows from this equation that the second derivative $\phi''(\cdot)$ is bounded, too, therefore the first derivative $\phi'(\cdot)$ of this solution is also bounded. Setting $\phi_0 = \phi(a) + \int_a^b \phi'(x)dx$ and $\phi_0' = \phi'(a) + \int_a^b \phi''(x)dx$ and considering the Cauchy problem for equation (II.1.1) with the initial data $\phi(b) = \phi_0$, $\phi'(b) = \phi_0'$, we immediately get our statement. In this section, we apply implicitly similar reasoning several times not specifying this in each case.

Let $f_1(\phi) = \overline{f}(\phi) - \omega\phi - a$ and $F_1(\phi) = \int_{a_-}^{\phi} f_1(p)dp$. Then, the equality

$$\{\frac{1}{2}(\phi')^2 + F_1(\phi)\}' = 0 \qquad\qquad (\text{II.1.2})$$

follows from (II.1.1). Suppose that $a_- = a_+ = a$, $F_1(b) = 0$, $f_1(b) > 0$ for some $b > a$ and $F_1(\phi) < 0$ if $\phi \in (a, b)$. Let us prove that if these assumptions are valid, then equation (II.1.1) has a solution satisfying the conditions $\phi(\pm\infty) = a$. Indeed, we take an arbitrary point $x_0 \in R$ and the following initial data for equation (II.1.1):

$$\phi(x_0) = b, \quad \phi'(x_0) = 0.$$

Then, by (I.1.1) $\phi''(x_0) < 0$, thus $\phi'(x) < 0$ in a right half-neighborhood of the point $x_0$. There cannot exist a point $x > x_0$ such that $\phi(x) = a$ because otherwise the function $E = \frac{1}{2}[\phi'(x)]^2 + F_1(\phi(x))$ (which we call the energy) must be positive at the point $x$ and, hence, non-equal to itself at $x_0$ ($\phi'(x) \neq 0$ for this point $x$ by the uniqueness theorem because $\phi \equiv a$ is a solution of the equation). Further, $\phi'(x) \neq 0$ for any $x > x_0$ such that $\phi(x) \in (a,b)$ because otherwise there is no conservation of the energy $E$. Hence $\phi(x) \in (a,b)$ for all $x > x_0$. The above facts easily imply, in particular, that the solution $\phi$ is global, i. e. it can be continued onto the whole half-line $x > x_0$. Also, in view of the above arguments, the graph of the function $\phi$ has a horizontal asymptote as $x \to +\infty$. A simple corollary of the above considerations is that $\phi(+\infty) = a$. By analogy, $\phi(-\infty) = a$, $\phi(x) \in (a,b)$ for all $x$ and $\phi'(x) > 0$ from the left of the point $x_0$. It can be easily proved that if $f_1'(a) = f(a) - \omega < 0$, then

$$|\phi(x) - a| + |\phi'(x)| \leq C_1 e^{-C_2 |x|}$$

for some positive constants $C_1$ and $C_2$. Thus, in this case $\phi - a \in H^1$.

By the complete analogy, if in equation (II.1.1) $f_1(a) = 0$ and $F_1(\phi) < 0$ from the right of $a$, then equation (II.1.1) possesses a solution $\phi$ tending to $a$ as $x \to +\infty$ (or as $x \to -\infty$), continuable on a half-line $x > d$ (resp. $x < d$) and decreasing (resp., increasing) on $(d, +\infty)$ (resp., on $(-\infty, d)$). If in addition $f_1(a_1) = F_1(a_1) = 0$ for some $a_1 > a$ and $F_1(\phi) < 0$ for $\phi \in (a, a_1)$, then it can be proved as above that this solution $\phi(x)$ is monotone on the entire real line and is such that $\phi(+\infty) = a$, $\phi(-\infty) = a_1$ (or conversely $\phi(-\infty) = a$, $\phi(+\infty) = a_1$). If there is no point $a_1 > a$ such that $F_1(a_1) = 0$ and consequently $F_1(\phi) < 0$ for $\phi > a$, then $\phi'(x) < 0$ (resp. $\phi'(x) > 0$) in all points where this solution exists and there is no finite limit $\lim_{x \to -\infty} \phi(x)$ (resp. $\lim_{x \to +\infty} \phi(x)$). Indeed, the first claim follows from (II.1.2) and the second is valid since if, for example, $\lim_{x \to -\infty} \phi(x) = a_1 > a$, then due to (II.1.2) it should be $F_1(a_1) = 0$ that is a contradiction. If the above point $a_1 > a$ satisfying $F_1(a_1) = 0$ and $F_1(\phi) < 0$ for $\phi \in (a, a_1)$ exists, then two cases are possible: if $f_1(a_1) = 0$, then, as above, $\phi'(x) < 0$ (resp. $\phi'(x) > 0$) for all $x \in R$ and $\lim_{x \to -\infty} \phi(x) = a_1$ (resp. $\lim_{x \to +\infty} \phi(x) = a_1$); if $f_1(a_1) > 0$, then we get the above solution $\phi(x)$ tending to $a$ as $|x| \to \infty$ and possessing precisely one point of maximum (since $f_1(a_1) = F_1'(a_1)$, always $f_1(a_1) \geq 0$).

An observation important for us following from the above arguments is that equation (II.1.1) can have bounded solutions possessing limits as $x \to \pm\infty$ of only two types. These are monotone solutions and solutions with precisely one point of extremum. We shall call the corresponding solitary waves *kinks* and *soliton-like solutions*, respectively.

Let us prove that the above-constructed solution $\phi = \phi(\omega, x)$ satisfying equation (II.1.1) and the conditions $\phi(\omega, \pm\infty) = a$, $\phi(\omega, x) > a$ for $x \in R$ and $\phi_x'(\omega, x_0) = 0$ is continuously differentiable as a function of the argument $\omega$ for an arbitrary $x \in R$ (in view of the invariance of the equation with respect to translations in

$x$, one cannot state the differentiability with respect to $\omega$ of an arbitrary family of solutions depending on the parameter $\omega$). It is sufficient to prove that the parameter $b$ introduced above is locally continuously differentiable as a function of $\omega \in (\omega_0 - \delta, \omega_0 + \delta)$ for some $\delta > 0$. But this follows from the implicit function theorem because $b$ is a solution of the algebraic equation $F_1(\omega, b) = 0$, where $\frac{\partial}{\partial r} F_1(\omega_0, r)\big|_{r=b} = \overline{f}(b) - \omega_0 b - a \neq 0$, and, thus, the continuous differentiability of the solution $\phi(\omega, x)$ with respect to $\omega$ is proved. Everywhere by $\phi'_\omega$ we denote the above function with a fixed point $x_0$.

Let $a = 0$, $f(0) - \omega_0 < 0$ and there exist $b > 0$ such that $F_1(b) = 0$, $f_1(b) = \overline{f}(b) - \omega_0 b > 0$ and $F_1(\phi) < 0$ for $\phi \in (0, b)$. Then, using the above methods, one can easily prove that $\phi'_\omega(\omega_0, x) \in H^1$ and

$$\frac{d}{d\omega} P(\phi)\Big|_{\omega=\omega_0} = 2 \int_{-\infty}^{\infty} \phi(\omega_0, x)\phi'_\omega(\omega_0, x)dx \quad \left( \text{here } P(\phi) = \int_{-\infty}^{\infty} \phi^2(\omega, x)dx \right).$$

If $f(\phi) = \phi^\nu$ and $\nu > 0$, then soliton-like solutions of the KdVE vanishing as $x \to \pm\infty$ are well known. They form the following one-parameter family

$$\phi = \left\{ A \text{ sech} \left[ \frac{A\nu}{\sqrt{(2(\nu+1)(\nu+2))}} \left( x - \frac{2A^2 t}{(\nu+1)(\nu+2)} \right) \right] \right\}^{\frac{2}{\nu}}, \tag{II.1.3}$$

where $\text{sech}(z) = \frac{2}{e^z + e^{-z}}$, $A > 0$ is a real parameter and $\omega = \frac{2A^2}{(\nu+1)(\nu+2)}$. One can easily verify this formula.

The above considerations show that the one-dimensional equation (II.1.1) relevant to the problem of the existence of solitary waves for the KdVE and NLSE with $N = 1$ is simple in a sense. Now, we begin to study the multi-dimensional case of the NLSE with $N \geq 2$. In this and the next sections, we consider radial solutions of the problem (II.0.1),(II.0.3). After the substitution $\phi(x) = y(r)$, where $r = |x|$, equation (II.0.1) takes the following form:

$$y'' + \frac{N-1}{r} y' = g(y), \quad r > 0, \tag{II.1.4}$$

where $g(y) = \omega y - f(y^2)y$ and the prime means the derivative in $r$. We supply this equation with initial data

$$y(0) = y_0 \in R, \quad y'(0) = 0. \tag{II.1.5}$$

First we obtain a result on the local well-posedness of this problem in a neighborhood of the point $r = 0$. The class of solutions $y(r)$ we admit consists of real-valued functions continuously differentiable in $[0, a)$ and twice continuously differentiable in $(0, a)$ where $0 < a \leq +\infty$.

<u>Theorem II.1.1</u> *Let* $g(\cdot) \in \bigcup\limits_{n=1}^{\infty} C^1((-n,n);R)$. *Then, for any* $y_0 \in R$ *there exists* $a > 0$ *such that the problem (II.1.4),(II.1.5) has a unique solution in* $[0,a)$. *This solution continuously depends on* $y_0$ *in the sense that, if for some* $\overline{y}_0 \in R$ *the corresponding solution* $\overline{y}(r)$ *can be continued on a segment* $[0,b]$, $b > 0$, *then solutions* $y(r)$ *of the problem (II.1.1),(II.1.2) for all* $y_0$ *from an interval* $(\overline{y}_0 - \delta, \overline{y}_0 + \delta)$, *where* $\delta > 0$, *can be continued on* $[0,b]$ *and the map* $y_0 \longmapsto y(r)$ *is continuous from* $(\overline{y}_0 - \delta, \overline{y}_0 + \delta)$ *into* $C^1([0,b];R)$.

Sketch of the proof. We shall consider only the case $N \geq 3$ because the case $N = 2$ can be studied by analogy. One can easily deduce by solving the equation that the linear problem with $h \in C([0,a);R)$

$$r^{1-N}(r^{N-1}w')' = h(r), \quad r > 0,$$

$$w(0) = w_0 \in R, \quad w'(0) = 0$$

in a neighborhood of zero has at most one solution of the above-described class. Further, the direct verification shows that an arbitrary solution of the class $C([0,a);R)$ of the nonlinear integral equation

$$y(r) = y_0 + (N-2)^{-1} \int\limits_0^r \left[1 - \left(\frac{s}{r}\right)^{N-2}\right] sg(y(s))ds \quad (r > 0), \quad y(0) = y_0 \quad \text{(II.1.6)}$$

satisfies the Cauchy problem (II.1.4),(II.1.5). Thus our Cauchy problem (II.1.4),(II.1.5) is equivalent to integral equation (II.1.6).

Let $M_a = \{u(r) \in C([0,a];R) : u(0) = y_0, |u(r) - y_0| \leq 1, r \in [0,a]\}$ where $a > 0$. Then, since for any $a > 0$ the kernel $k(r,s) = (N-2)^{-1}\left[1 - \left(\frac{s}{r}\right)^{N-2}\right]s$ of the integral in equation (II.1.6) is a continuous and bounded function of $(r,s) \in (0,a] \times [0,a]$, one can easily get by standard methods that, for a sufficiently small $a > 0$, the operator from the right-hand side of (II.1.6) maps the set $M_a$ into itself and is a contraction. Thus, it has a fixed point $y(r) \in M_a$. So, we have proved the unique local solvability of the problem (II.1.4),(II.1.5). The local continuous dependence of solutions of this problem on the parameter $y_0$ (in a sufficiently small neighborhood of zero) follows from equation (II.1.6) by standard methods based on the Gronwell's lemma. After that the global continuous dependence is a corollary of the regularity of equation (II.1.4) from outside of any neighborhood of zero.$\square$

Here we shall illustrate with a model example the ODE approach, i. e. the approach this section is devoted to. We shall prove the following.

<u>Theorem II.1.2</u> *Let* $g$ *be a continuously differentiable function,* $g(a_{-1}) = g(0) = g(a_1) = g(a_2) = 0$ *for some* $a_{-1} < 0 < a_1 < a_2$, $g(y) > 0$ *if* $y \in (a_{-2}, a_{-1}) \bigcup (0, a_1)$

*for some $a_{-2} < a_{-1}$, $g(y) < 0$ if $y \in (a_{-1}, 0) \bigcup (a_1, a_2)$ and $G(a_{-2}) \geq G(a_2) > 0$ where $G(s) = -2 \int_0^s g(r) dr$. Then for any $l = 0, 1, 2, 3, \ldots$ the problem (II.1.4),(II.1.5) has a solution continuable on the half-line $r > 0$, tending to zero as $x \to +\infty$ and possessing precisely $l$ roots in the half-line $(0, \infty)$.*

Our <u>Proof</u> of this theorem consists of two steps. At the first step we show that the Cauchy problem (II.1.4),(II.1.5) has solutions, for which $y_0 \in (a_1, a_2)$, with an arbitrary large number of roots in the half-line $r > 0$. At the second step, we take for $y_0$ the greatest lower bound of its values from $(a_1, a_2)$ for each of which the solution of the Cauchy problem (II.1.4),(II.1.5) has no less than $(l + 1)$ roots, where $l \geq 0$ is integer. We show that the corresponding solution of the Cauchy problem has exactly $l$ roots and tends to zero as $r \to +\infty$.

Let us sketch the first part of the proof. With this we follow the paper [98]. Take an arbitrary integer $l \geq 0$. For an arbitrary solution $y(r)$ of the problem (II.1.4),(II.1.5) the following identity takes place:

$$\{y'^2 + G(y)\}' = -\frac{2(N-1)}{r} y'^2. \tag{II.1.7}$$

Hence, since $G(y) < G(a_2) \leq G(a_{-2})$ for $y \in (0, a_2)$ and the function $[y'(r)]^2 + G(y(r))$ of the argument $r > 0$ is nonincreasing, for any $y_0 \in (a_1, a_2)$ the solution $y(r)$ of the problem (II.1.4),(II.1.5) satisfies the estimate

$$a_{-2} \leq y(r) \leq a_2, \quad r > 0, \tag{II.1.8}$$

and correspondingly it can be continued on the entire half-line $r > 0$. In addition, clearly due to (II.1.7) there exists $C_0 > 0$ such that

$$|y'(r)| \leq C_0, \quad r > 0, \tag{II.1.9}$$

for any of these solutions. We call the function $E(r) = (y'(r))^2 + G(y(r))$ the energy of a solution $y(r)$.

Since $y(r) \equiv a_2$ is a solution of equation (II.1.4), taking $y_0$ from the interval $(a_1, a_2)$ sufficiently close to $a_2$, we get by the theorem on continuous dependence that there exist $y_0 \in (a_1, a_2)$ and a sufficiently large $r_0 > 0$ such that

$$G(y(r_0)) > \frac{2C_0(a_2 - a_{-2})(l + 2)(N-1)}{r_0} \tag{II.1.10}$$

and $y(r) \in (a_1, a_2)$ for all $r \in [0, r_0]$. Let $r_1$ and $r_2$ be two points such that $y'(r) \neq 0$ for $r \in (r_1, r_2)$. Then, it follows from (II.1.8),(II.1.9) that

$$2(N-1) \int_{r_1}^{r_2} \frac{y'(r)^2}{r} dr \leq \frac{2(N-1)C_0(a_2 - a_{-2})}{r_1}.$$

Now, let $y(r)$ be a solution of the Cauchy problem (II.1.4),(II.1.5) satisfying (II.1.10) and let $r_1$ be the root of $y'(r)$ on the right of $r_0$ nearest to $r_0$ or $r_1 = +\infty$ if there are no roots of $y'(r)$ on the right of $r_0$. Then, according to the latter inequality and inequality (II.1.10), we have:

$$E(r_1) \geq \frac{2C_0(a_2 - a_{-2})(l+1)(N-1)}{r_0}.$$

If $r_1 = +\infty$, then our solution is monotone for $r > r_0$. Therefore, its graph has an asymptote $y = c \in (a_{-2}, a_2)$. Since $f(y) = 0$ in the interval $(a_{-2}, a_2)$ only if $y = a_{-1}, 0, a_1$, one has that $c$ is equal to one of these three numbers. But this contradicts the positiveness of the energy for all $r > r_0$ because $G(a_{-1}), G(0), G(a_1) \leq 0$. Therefore, there exists a root $r_1 > r_0$ of $y'(r)$. We assume that $r_1 > r_0$ is the root of $y'(r)$ nearest to $r_0$. Since $E(r_1) > 0$ and due to the maximum principle (any solution can have only a point of maximum in the domain $(a_{-1}, 0) \bigcup (a_1, a_2)$ and only a point of minimum in the domain $(a_{-2}, a_{-1}) \bigcup (0, a_1)$), we have $y(r_1) \in (a_{-2}, a_{-1})$. Hence, our solution $y$ has a root greater than $r_0$. Repeating these arguments (i. e. changing $r_0$ by $r_1$ and so on), we get that our solution $y(r)$ has no less than $(l+1)$ roots on the right of $r_0$, and the first part of Theorem II.1.2 is proved.

Now, we turn to the second part of our proving. In this part of the proof, we follow the paper [110]. First of all, we make two remarks. First, since as it was noted earlier, if $y(r_0) = 0$ for a solution $y(r)$ of the problem (II.1.4),(II.1.5) and a point $r_0 > 0$, then $y'(r_0) \neq 0$, we have that roots of an arbitrary solution of our problem are isolated. Second, we show that between any two nearest roots of a nontrivial solution $y(r)$ with $y(0) \in (a_1, a_2)$ lies a unique point of extremum of this solution. Indeed, let $r_1 < r_2$ be two nearest positive roots of a solution $y(r)$. Let also for the definiteness $y(r) > 0$ between $r_1$ and $r_2$. Then, $y'(r_1) > 0$, hence there exists $\bar{r} \in (r_1, r_2)$ such that $y'(\bar{r}) = 0$ and $y'(r) > 0$ for all $r \in (r_1, \bar{r})$. Then, by (II.1.7) $y(\bar{r}) \in (a_1, a_2)$ (otherwise $E(\bar{r}) \leq 0$). Therefore, $y''(\bar{r}) < 0$, hence $y'(r) < 0$ in a right half-neighborhood of $\bar{r}$. Suppose the existence of $\bar{\bar{r}} \in (\bar{r}, r_2)$ such that $y'(\bar{\bar{r}}) = 0$ and $y'(r) < 0$ for $r \in (\bar{r}, \bar{\bar{r}})$. Then, according to equation (II.1.4) $y(\bar{\bar{r}}) \in (0, a_1]$ because if $y(\bar{\bar{r}}) \in (a_1, a_2)$, then $y''(\bar{\bar{r}}) < 0$ which contradicts the facts that $y'(r) < 0$ in a left half-neighborhood of $\bar{\bar{r}}$ and that $y'(\bar{\bar{r}}) = 0$. But then $E(\bar{\bar{r}}) \leq 0$, and we get a contradiction because $E(r_2) > 0$ where $r_2 > \bar{\bar{r}}$. So, our second claim is proved.

Let us take an arbitrary integer $l \geq 0$, let

$$Y_l = \{y_0 \in (0, a_2) : \ y(r) \text{ has no less than } (l+1) \text{ roots}\}$$

and $\bar{y}_0 = \inf Y_l$. By $\bar{y}$ we denote the solution of the problem (II,1.4),(II.1.5) with $y_0 = \bar{y}_0$. First, $\bar{y}_0 > 0$ because due to (II.1.7) a solution $y(r)$ does not have roots if $y_0 \in (0, a_1]$ since $G(y_0) < 0$ and consequently $E(0) < 0$ for these values of $y_0$. Let

$\{y_0^m\}$ be a sequence from $Y_l$ converging to $\bar{y}_0$. Now we want to prove that the smallest $l$ roots $r_1^m < ... < r_l^m$ of the solution $y^m$ of the Cauchy problem (II.1.4),(II.1.5) taken with $y_0 = y_0^m$ are bounded uniformly with respect to $m = 1, 2, 3, ....$ Let $\bar{r}_1^m < ... < \bar{r}_{l+1}^m$ be the smallest points of extremum of $y^m$. Then, as it is proved above, $0 = \bar{r}_1^m < r_1^m < \bar{r}_2^m < ... < r_l^m < \bar{r}_{l+1}^m$.

<u>Lemma II.1.3</u> *There exist constants $a, b > 0$ such that*

$$|\bar{r}_{k-1}^m - r_k^m| \le a + b(\bar{r}_{k+1}^m)^{\frac{1}{2}}, \quad |r_k^m - \bar{r}_{k+1}^m| \le a + b(\bar{r}_{k+1}^m)^{\frac{1}{2}} \quad (k = 1, ..., l)$$

*for all $m$.*

<u>Proof.</u> We will prove only the second inequality because the first can be proved by analogy. Suppose for the definiteness that $y^m(\bar{r}_{k+1}^m) > 0$. Let $z \in (r_k^m, \bar{r}_{k+1}^m)$ be such that $y^m(z) = d$ where $d$ is the point from the interval $(a_1, a_2)$ for which $G(d) = \frac{G(a_1)}{2} < 0$. First, we prove that

$$|\frac{d}{dr}y^m(r)| \ge C(\bar{r}_{k+1}^m)^{-\frac{1}{2}} \tag{II.1.11}$$

for all $r \in (r_k^m, z)$ where $C > 0$ is independent of $m$. We have $G(y^m(\bar{r}_{k+1}^m)) > 0$; therefore, by (II.1.7) for any $r \in (z, z_1)$ one has $0 < C_1 \le |y^{m\prime}(r)| \le C_0$ for some $C_1$ independent of $m$ where $z_1 \in (z, \bar{r}_{k+1}^m)$ is such that $y^m(z_1) \in (a_1, a_2)$ and $G(y^m(z_1)) = \frac{G(a_1)}{4}$ (of course, $z$ and $z_1$ depend on $m$ but we omit this index for the simplicity of the notation). Hence

$$y'^2(r) \ge 2(N-1) \int_z^{\bar{r}_{k+1}^m} \frac{y'^2(r)}{r}dr \ge C_2(\bar{r}_{k+1}^m)^{-1}$$

if $r \in (r_k^m, z)$. Thus, (II.1.11) holds. Finally, by (II.1.4) we have $\frac{d^2}{dr^2}y^m(r) \le -C_3 < 0$ for all $r \in (r_k^m, \bar{r}_{k+1}^m)$ such that $y^m(r) \ge d$. In view of this fact and (II.1.11) we get

$$|r_k^m - \bar{r}_{k+1}^m| \le a + b(\bar{r}_{k+1}^m)^{\frac{1}{2}},$$

and Lemma II.1.3 is proved.□

Summing the inequalities from the statement of Lemma II.1.3, we obtain

$$\bar{r}_{l+1}^m \le C_1(\bar{r}_{l+1}^m)^{\frac{1}{2}} + C_2.$$

Hence, $\{\bar{r}_{l+1}^m\}_{m=1,2,3,...}$ is a bounded sequence. By the estimate (II.1.9) and since the values of the functions $y^m$ at their first $l+1$ points of extremum lie from the outside of the interval $(a_{-1}, a_1)$, there exists a constant $C > 0$ such that

$$|r_k^m - r_{k-1}^m| \ge C, \quad k = 1, 2, ..., l, \quad m = 1, 2, 3, ....$$

Thus, the function $\bar{y}$ has no less than $l$ roots. At the same time, the theorem on the continuous dependence of solutions on the parameter $y_0$ and the fact that $\bar{y}'(r_0) \neq 0$ if $y(r_0) = 0$ imply in view of the definition of $\bar{y}_0$ that the function $\bar{y}$ cannot have more than $l$ roots.

Let us prove that $\lim_{r \to \infty} \bar{y}(r) = 0$. Suppose this is not the case. Then, either the solution $\bar{y}$ is monotone for sufficiently large $r$ or it has a sequence of extrema $z_n \to +\infty$. In the first case, the graph of this function has an asymptote which can be only $y = a_{-1}$ or $y = a_1$. Hence, the energy $\overline{E}(r)$ of the solution $\bar{y}(r)$ is negative for sufficiently large values of the argument $r$. In the second case, since the number of roots of the solution $\bar{y}$ is equal to $l$, it should have extrema in the domain $y \in (a_{-1}, 0) \bigcup (0, a_1)$. Thus, in this case the energy $\overline{E}(r)$ is negative for sufficiently large values of $r$, too. But the negativeness of the energy implies, by the theorem on continuous dependence of solutions of the problem (II.1.4),(II.1.5) on the parameter $y_0$, that for sufficiently large values $m$ solutions $y^m$ cannot have more than $l$ roots. This contradiction implies that $\lim_{r \to \infty} \bar{y}(r) = 0$, and Theorem II.1.3 is completely proved.□

## 2.2   Existence of solutions.  A variational method

In this section we shall consider an application of variational methods to proving the existence of radial solutions of the problem (II.0.1),(II.0.3). We restrict our attention to the case $f(\phi^2) = |\phi|^{p-1}$, $p > 1$. So, we consider the problem

$$\Delta \phi = \omega \phi - |\phi|^{p-1}\phi, \quad x \in R^N, \tag{II.2.1}$$

$$\phi|_{|x| \to \infty} = 0. \tag{II.2.2}$$

Our result on the existence is the following.

Theorem II.2.1 Let $\omega > 0$, $N \geq 3$ be integer, and $p \in (1, \frac{N+2}{N-2})$. Then, the problem (II.2.1),(II.2.2) has a positive radial solution and, for any $l = 1, 2, 3, ...$, a radial solution $u_l = u_l(r)$, where $r = |x|$, with precisely $l$ roots on the half-line $r > 0$. We note that a similar result takes place for the problem (II.2.1),(II.2.2) with the right-hand side of the equation of a more general kind $g(\phi)$ for functions $g(\phi)$ in a sense similar to $\omega \phi - |\phi|^{p-1}\phi$.

Two results which we present below are used when proving Theorem II.2.1.

Theorem II.2.2 Let $H$ be a real Hilbert space with a norm $|| \cdot ||$, $J$ be a continuously differentiable real-valued functional in $H$, and $S = \{h \in H : ||h|| = 1\}$. Let $r(\cdot) > 0$ be a continuously differentiable function on $S$ such that for any $v \in S$ $J'_r(rv)\big|_{r=r(v)} = 0$. Then, if the functional $\hat{J}(v) = J(r(v)v)$ considered on $S$ has a critical point $v_0 \in S$, then $J'(h)\big|_{h=r(v_0)v_0} = 0$.

<u>Proof</u> is clear. By conditions of the theorem

$$< J'(r(v_0)v_0), v_0 > = \frac{d}{dr} J(rv_0)\big|_{r=r(v_0)} = 0.$$

Further, consider an arbitrary continuously differentiable map $\gamma$ from $(-1,1)$ into $H$ such that $\gamma(s) \in S$ for $s \in (-1,1)$, $\gamma(0) = v_0$ and $\gamma'(0) \neq 0$. Then,

$$0 = \frac{d}{ds}\hat{J}(\gamma(s))\big|_{s=0} = < J'(r(v_0)v_0), \frac{d}{ds}[r(\gamma(s))]\big|_{s=0}v_0 + r(v_0)\gamma'(0) > =$$

$$= r(v_0) < J'(r(v_0)v_0), \gamma'(0) > .$$

The set of all vectors of the kind $\gamma'(s)\big|_{s=0}$ is obviously the subspace $L$ of the space $H$ consisting of all vectors orthogonal to $v_0$, therefore $< J'(h)\big|_{h=r(v_0)v_0}, w > = 0$ for any $w \in L$. Thus, $< J'(r(v_0)v_0), w > = 0$ for any $w \in H$.□

<u>Remark II.2.3</u> Theorem II.2.2 is a simplification, sufficient for our goals, of results by S.I. Pohozaev from [75,76].

The second result we need is the following.

<u>Theorem II.2.4</u> *Let $N \geq 3$ and $H_r^1 = \{h \in H^1 : h = h(|x|)\}$ be the subspace of $H^1$ with the scalar product and the norm of $H^1$. Then, the embedding of $H_r^1$ into $L_q$ is compact for $2 < q < \frac{2N}{N-2}$; also, for an arbitrary $h \in H_r^1$ there exists a unique $\hat{h} \in H_r^1$ coinciding with $h$ almost everywhere in $R^N$ and continuous everywhere except $x = 0$. In addition, $\hat{h}(|x|) \to 0$ as $|x| \to \infty$. Thus, we can accept that each element of $H_r^1$ is a function continuous at any point $x \neq 0$ and vanishing as $|x| \to \infty$.*

Sketch of the Proof. Clearly, the set $C_{0,r}^\infty$ of all radial functions from $C_0^\infty$ is dense in $H_r^1$. For any $g \in C_{0,r}^\infty$

$$\|g\|_{1,r} = D_N \int_0^\infty r^{N-1}[g^2(r) + (g'(r))^2]dr$$

where the constant $D_N > 0$ depends only on $N$. Hence the space $H_r^1$ is a completion of the linear space $C_{0,r}^\infty$ equipped with the norm $\|\cdot\|_{1,r}$. Let $h \in H_r^1$ and $\{g_n\}_{n=1,2,3,\ldots} \subset C_{0,r}^\infty$ be a sequence converging to $h$ in $H_r^1$. Then obviously for any $r > 0$ there exists

$$\hat{h}(r) = -\lim_{n\to\infty} \int_r^\infty g_n'(s)ds \qquad (II.2.3)$$

and for any $a, b : 0 < a < b$   $g_n \to \hat{h}$ as $n \to \infty$ in $H^1(a,b)$; in particular $\hat{h}(\cdot)$ is continuous on the half-line $r > 0$. Further, since the sequence $\{g_n\}_{n=1,2,3,\ldots}$ is converging in $H^1_r$, we have $\hat{h} = \lim\limits_{n\to\infty} g_n \in H^1_r$ and thus, as an element of the space $H^1_r$, $\hat{h}$ coincides with $h$, therefore $\hat{h}(|x|) = h(|x|)$ almost everywhere in $R^N$. These facts and (II.2.3) easily imply that

$$\hat{h}(r) = -\int_r^\infty \hat{h}'(s)ds, \quad r > 0,$$

hence $\hat{h}(+\infty) = 0$. Also, we obviously have

$$|\hat{h}(r)| = \left|\int_r^\infty h'(s)ds\right| \leq \left\{\int_r^\infty s^{N-1}(h'(s))^2 ds\right\}^{\frac{1}{2}} \left\{\int_r^\infty \frac{ds}{s^{N-1}}\right\}^{\frac{1}{2}} \leq (N-2)^{-\frac{1}{2}}||h||_1 r^{\frac{2-N}{2}}.$$

$$(II.2.4)$$

Let $\{h_n\}_{n=1,2,3,\dots} \subset H^1_r$ be a bounded sequence of functions continuous in $R^N \setminus \{0\}$. Each $h_n$ belongs to $L_q$ by the embedding of $H^1$ into $L_q$. Further, for $r > 0$ by (II.2.4) $\max\limits_{r\geq R}|h_n(r)| \leq C\epsilon(R)$ where $C > 0$ and $\epsilon(R)$ are independent of $n$ and $\epsilon(R) \to 0$ as $R \to +\infty$. Hence for $R > 0$

$$|h_n|^q_{L_q(\{x:\ |x|>R\})} = D_N \int_R^\infty r^{N-1}|h_n(r)|^q dr \leq D_N C^{q-2}\epsilon^{q-2}(R)||h_n||^2_{1,r}.$$

Since here for an arbitrary given $\delta > 0$ the right-hand side is smaller than $\delta$ for sufficiently large $R > 0$ and all $n = 1, 2, 3, \dots$, the sequence $\{h_n\}_{n=1,2,3,\dots}$ is compact in $L_q$. $\square$

Remark II.2.5 Theorem II.2.4 is a particular case of embedding theorems of Sobolev spaces of functions with symmetries obtained by P.L. Lions in [56].

Now, we turn to proving Theorem II.2.1. We consider the following spaces of functions: the space $H^1_r(a,b)$ of functions $u(|x|)$ from $H^1_r$ satisfying the condition $u(|x|) = 0$ as $|x| \leq a$ or $|x| \geq b$ (here $0 < a < b$) and the space $H^1_r(b)$ of functions from $H^1_r$ satisfying $u(|x|) = 0$ as $|x| \geq b$. Let $H^1_r(0,b) = H^1_r(b)$ and $H^1_r(+\infty) = H^1_r$. Clearly $u \in H^1_r$ for any $u \in H^1_r(a,b)$.

Consider the functional

$$J(\phi) = \int_{R^N} \{\frac{1}{2}(|\nabla\phi|^2 + \omega\phi^2) - \frac{1}{p+1}|\phi|^{p+1}\}dx.$$

It is clear that $J$ is continuously differentiable in $H^1_r$.

Consider arbitrary $a, b:\ 0 \leq a < b \leq +\infty$. First, let us prove that the problem

$$\Delta\phi = \omega\phi - |\phi|^{p-1}\phi, \quad |x| \in (a,b)\ (|x| < b\text{ if } a = 0), \tag{II.2.5}$$

$$\phi(a) = \phi(b) = 0 \quad \text{(to be interpreted as } \phi(b) = 0 \text{ if } a = 0), \tag{II.2.6}$$

has a positive solution. We use Theorem II.2.2. Let $S = \{h \in H_r^1 : \|h\|_1 = 1\}$ and $v \in S \cap H_r^1(a, b)$. Then, the conditions of Theorem II.2.2 $r = r(v) > 0$ and $J_r'(rv)\big|_{r=r(v)} = 0$ imply for our case

$$r(v) = \left\{ \frac{\displaystyle\int_{a<|x|<b} (|\nabla u|^2 + \omega u^2)dx}{\displaystyle\int_{a<|x|<b} |u|^{p+1}dx} \right\}^{\frac{1}{p-1}}$$

and, since $\frac{d}{dr}J(rv)\big|_{r=r(v)} = 0$, functions $\phi = r(v)v$ satisfy the identity

$$\int_{a<|x|<b} (|\nabla \phi|^2 + \omega \phi^2)dx = \int_{a<|x|<b} |\phi|^{p+1}dx. \tag{II.2.7}$$

Obviously, $r(v)$ is a smooth function of $v \in S \cap H_r^1(a, b)$. We set

$$M = \{\phi \in H_r^1(a, b) : \ \phi \neq 0 \text{ and } \phi \text{ satisfies (II.2.7)}\}.$$

Then we get

$$J(\phi) = (\frac{1}{2} - \frac{1}{p+1})\|\phi\|_1^2 \quad \text{for} \ \ \phi \in M; \tag{II.2.8}$$

therefore, the functional $J$ is bounded from below on $M$. In addition, since by the embedding theorem and (II.2.7) $\|\phi\|_1^2 \leq C\|\phi\|_1^{p+1}$ with $C > 0$ independent of $\phi \in M$, there exists $C_1 > 0$ such that

$$\|\phi\|_1 \geq C_1 > 0 \tag{II.2.9}$$

for all $\phi \in M$.

Let $\{\phi_n\}$ be a minimizing sequence for the functional $J$ on $M$. In view of (II.2.8), this sequence is bounded in $H_r^1$; hence, it is weakly compact. Let $\{\phi_{n_m}\}$ be a subsequence weakly converging to some $\phi$ in $H_r^1$. We can also accept that there exists $\lim_{m\to\infty} \|\phi_{n_m}\|_1$. Then, $\|\phi\|_1 \leq \lim_{m\to\infty} \|\phi_{n_m}\|_1$. Suppose that $\|\phi\|_1 < \lim_{m\to\infty} \|\phi_{n_m}\|_1$. By Theorem II.2.4 and by (II.2.7),(II.2.9) we have:

$$\lim_{m\to\infty} \int_{R^N} |\phi_{n_m}|^{p+1}dx = \int_{R^N} |\phi|^{p+1}dx \neq 0.$$

Hence, for our $\phi$, the left-hand side of (II.2.7) is smaller than the right-hand side and the latter is positive. Therefore, $\phi \neq 0$ and there exists $\alpha \in (0, 1)$ such that $\alpha\phi \in M$. But for this reason by (II.2.8) we get:

$$J(\alpha\phi) = \alpha^2(\frac{1}{2} - \frac{1}{p+1})\|\phi\|_1^2 < \lim_{m\to\infty} (\frac{1}{2} - \frac{1}{p+1})\|\phi_{n_m}\|_1^2 = \inf_{v\in M} J(v),$$

i.e. we get a contradiction. Hence, $\lim\limits_{m\to\infty} \|\phi_{n_m}\|_1^2 = \|\phi\|_1^2$, $\phi \in M$ and $J(\phi) = \inf\limits_{v\in M} J(v)$.

Further, $|\phi| \in M$, too, and $J(|\phi|) = J(\phi)$. Hence, changing $\phi$ by $|\phi|$, we can accept that $\phi \geq 0$. According to Theorem II.2.2, $\phi$ is a critical point of the functional $J$ considered on the space $H_r^1(a, b)$. Taking $\frac{d}{dt}J(\phi + tv)|_{t=0}$ for an arbitrary radial function $v(r) \in C_0^\infty(a, b)$ satisfying $v(|x|) = 0$ in a neighborhood of zero, we obtain the equality

$$D_N \int\limits_0^{+\infty} \left\{r^{N-1}\left(\frac{d}{dr}v(r)\frac{d}{dr}\phi(r) + \omega v(r)\phi(r) - v(r)\phi^p(r)\right)\right\}dr = 0, \qquad \text{(II.2.10)}$$

hence, $\phi$ is a generalized solution of equation (II.2.5) in any bounded domain contained with its closure in the set $\{x \in R^N : a < |x| < b\}$. If $a > 0$, then $\phi(a) = \phi(b) = 0$ and (II.2.10) implies that the function $\phi(|x|)$ satisfies equation (II.2.5) in the classical sense in the domain $\{x \in R^N : a < |x| < b\}$. Hence, in particular, $\phi(|x|) > 0$ for $a < |x| < b$ (because otherwise $\phi(r) = \phi_r'(r) = 0$ for some $r \in (a, b)$ which is impossible because of the uniqueness theorem, see Section 2.1).

Consider the case $a = 0$. Then, the equality

$$\int\limits_{R^N} \{\nabla v(x)\,\nabla\phi(x) + \omega v(x)\phi(x) - v(x)\phi^p(x)\}dx = 0 \qquad \text{(II.2.11)}$$

also follows from the above arguments for non-radial functions $v \in C_0^\infty$ satisfying the condition $v(x) = 0$ for $|x| \geq b$ and for $x$ from an open neighborhood of zero. Taking here the limit over a sequence $\{v_n\}$ satisfying these properties and converging to a function $v \in C_0^\infty$ in the sense of the space $H^1$, we achieve equality (II.2.11) for an arbitrary $v \in C_0^\infty$ with a support in the domain $|x| \leq b$. Hence, $\phi$ is a generalized solution of elliptic equation (II.2.5) in the domain $|x| < b$ which is also a radial function. If $b = +\infty$, then $\phi|_{|x|\to\infty} = 0$ by Theorem II.2.4. Also, since the function $\phi$ satisfies equation (II.1.4) with $g(u) = \omega u - |u|^{p-1}u$, it cannot be equal to zero in a point of the domain $a < |x| < b$ as in the previous section.

Let us prove that the solution $\phi(|x|)$ of equation (II.2.5) is classical, i. e. that it is twice continuously differentiable and satisfies this equation at any point $x : |x| < b$. Clearly, it suffices to prove this fact in an arbitrary small neighborhood of the point $x = 0$. We take an arbitrary $R > 0$ if $b = +\infty$ and $R = b$ if $b < +\infty$. Then, since $\phi \in W_2^1(B_R(0))$, we have $\phi \in L_{q_0}(B_R(0))$ where $q_0 = \frac{2N}{N-2}$. Hence, $|\phi|^{p-1}\phi \in L_{r_1}(B_R(0))$ with $r_1 = \psi(q_0) = \frac{q_0}{p} = \frac{2N}{p(N-2)}$. Therefore, $\phi \in W_{r_1}^2(B_R(0))$ (on this subject see [36]). Thus, by the embedding theorem $\phi \in L_{q_1}(B_R(0))$, where $q_1 = \varphi(r_1) = \frac{r_1 N}{N-2r_1} = \frac{2N}{p(N-2)-4}$ if $p \in \left(\frac{4}{N-2}, \frac{N+2}{N-2}\right)$ and $q_1 > 1$ is arbitrary if $p = \frac{4}{N-2}$; $\phi \in C(\overline{B_R(0)})$ if $p \in \left(1, \frac{4}{N-2}\right)$. In the two latter cases $\phi \in W_r^2(B_R(0))$ with arbitrary large $r > 1$, hence, $\phi$ is continuously differentiable in $B_R(0)$, therefore it is a classical solution of our equation.

Consider the case $p \in \left(\frac{4}{N-2}, \frac{N+2}{N-2}\right)$. One can easily verify that $q_1 > q_0$. We also observe that $q_1 = \varphi(\psi(q_0))$. By analogy, one can show that, if $r_2 = \psi(q_1) = \frac{q_1}{p} < \frac{N}{2}$, then $\phi \in L_{q_2}(B_R(0))$ with $q_2 = \varphi(r_2) > q_1$, if $r_3 = \psi(q_2) = \frac{q_2}{p} < \frac{N}{2}$, then $\phi \in L_{q_3}(B_R(0))$ with $q_3 = \varphi(r_3) > q_2$, and so on. So, we get either infinite increasing sequences $\{r_n\}_{n=1,2,3,\ldots}$ and $\{q_n\}_{n=0,1,2,\ldots}$ satisfying $r_n < \frac{N}{2}$ for all $n$ or a number $n_0$ such that $r_1 < r_2 < \ldots < r_{n_0-1} < \frac{N}{2}$ and $r_{n_0} \geq \frac{N}{2}$. First of all, in the case when these sequences are infinite it cannot be that $r_n \leq \frac{N}{2} - \delta$ for some $\delta > 0$ and all $n$ because the sequence $\{q_n\}_{n=0,1,2,\ldots}$ is monotone and bounded, hence, it must converge to a fixed point of the map $\varphi(\psi(\cdot))$, but it can be simply verified that this map has no fixed point greater than $q_0 = \frac{2N}{N-2}$. In the remaining cases by the embedding theorems $\phi$ belongs to $L_q(B_R(0))$ with arbitrary large $q > 1$, therefore by embedding theorems it is continuously differentiable, thus, it is a classical solution of equation (II.2.5) and the required statement is proved.

Let us fix an arbitrary positive integer $l$. Let $0 < a_1 < a_2 < \ldots < a_l < \infty$ be arbitrary points and let $a_0 = 0$ and $a_{l+1} = +\infty$. Let $u_1(|x|), u_2(|x|), \ldots, u_{l+1}(|x|)$ be the above-constructed radial solutions of the problem (II.2.5),(II.2.6) with $a = a_{k-1}$ and $b = a_k$ (i.e. they satisfy the above minimization problems), respectively, where $u_1 > 0, u_3 > 0, \ldots$ and $u_2 < 0, u_4 < 0, \ldots$ (solutions $u_2, u_4, \ldots$ exist because if $u$ is a positive solution of the problem (II.2.5),(II.2.6), then $-u$ is a negative solution). We denote by $u(|x|)$ the function which is equal to $u_k(|x|)$ in the domain $a_{k-1} \leq |x| \leq a_k$, $k = 1, 2, \ldots, l+1$. Let $R$ be the set of all these functions for all values of the parameters $a_k$. Then, according to (II.2.8) $J(u) = \left(\frac{1}{2} - \frac{1}{p+1}\right)\|u\|_1^2$ for any $u \in R$ and for all values of $a_k$, i.e., the functional $J$ is bounded from below on the set $R$. Let $\{\phi_n\}_{n=1,2,3,\ldots}$ be a minimizing sequence for this functional $J$ considered on the set $R$ with the corresponding values of parameters $0 = a_0^n < a_1^n < a_2^n < \ldots < a_l^n < a_{l+1}^n = +\infty$ ($n = 1, 2, 3, \ldots$).

<u>Lemma II.2.6</u> *There exists a subsequence of the sequence $a^n = (a_1^n, \ldots, a_l^n) \in R^l$, $n = 1, 2, 3, \ldots$, converging to some $a = (a_1, \ldots, a_l) \in R^l$ where $0 < a_1 < a_2 < \ldots < a_l < +\infty$.*

<u>Proof.</u> It suffices to prove that there exist $C, C_1 > 0$ such that $a_{k+1}^n - a_k^n \geq C$, $k = 0, \ldots, l-1$, and $a_l^n \leq C_1$, for all $n$. We shall only prove that $a_1^n \geq C > 0$ ($n = 1, 2, 3, \ldots$) because the other estimates can be obtained by analogy.

It suffices to prove that

$$\inf_{u \in H_0^1(b) \cap M, \ u \neq 0} J(u) \to +\infty$$

as $b \to +0$. Suppose it is not right. Then, there exists a sequence $b_n \to +0$ such that $\|\phi_n\|_1 \leq C_2$ for some minimizers $\phi_n$ (here $n = 1, 2, 3, \ldots$). Then $|\phi_n|_{p+1} \leq C_2'$, too. Let $v_n(x) = \alpha_n \phi_n(b_n|x|)$ where $\alpha_n > 0$ are chosen for the functions $v_n$ to satisfy condition (II.2.8) with $a = 0$ and $b = 1$. Then, using embedding theorems, we get $||v_n||_1^2 \leq C_3 |\nabla v_n|_2^2$, therefore

$$b_n^2 |\nabla \phi_n|_2^2 \geq C_4 \alpha_n^{p-1} |\phi_n|_{p+1}^{p+1}$$

for some $C_3, C_4 > 0$ independent of $n$. Hence, since as earlier $|\phi_n|_{p+1} \geq C_5 > 0$ for all $\phi \in H_r^1$ satisfying (II.2.8), we have

$$0 < \alpha_n \leq C_6 b_n^{\frac{2}{p-1}} \to +0 \quad \text{as} \quad n \to \infty.$$

Therefore,

$$J(v_n) = \left( \frac{1}{2} - \frac{1}{p+1} \right) ||v_n||_1^2 \leq C_7 |\nabla v_n|_2^2 \leq C_8 b_n^{2-N+\frac{4}{p-1}} |\nabla \phi_n|_2^2 \leq C_9 b_n^{2-N+\frac{4}{p-1}} \to 0$$

as $n \to \infty$ because $2 - N + \frac{4}{p-1} > 0$. In particular, this implies that

$$\inf_{v \in H_0^1(1) \cap M, \ v \neq 0} J(v) \leq 0,$$

i. e. a contradiction.□

Let us return to the proof of Theorem II.2.1. As when proving the existence of positive solutions, one can prove, passing to a subsequence if necessary, that the sequence $\{\phi_n\}_{n=1,2,3,...}$ weakly converges in $H_r^1$ to a function $\phi(|x|)$ which satisfies condition (II.2.8) with $a = a_k$ and $b = a_{k+1}$, $k = 0, 1, ..., l$, and equation (II.2.5) for $a_k < |x| < a_{k+1}$ $(k = 0, 1, ..., l)$; in addition $\phi(a_k) = 0$ for $k = 1, 2, ..., l$. Also, since the function $\phi(r)$, where $r = |x|$, is a solution of equation (II.1.4) for $r \neq a_k$, there exist a derivative on the right $\phi'(a_k + 0)$ and a derivative on the left $\phi'(a_k - 0)$ which are nonequal to zero by the uniqueness theorem.

Now, to prove Theorem II.2.1, it suffices to show that $\phi'(a_k - 0) = \phi'(a_k + 0)$ for all $k = 1, 2, ..., l$. Let us take an arbitrary $k = 1, 2, ..., l$ and suppose that $\phi'(a_k - 0) \neq \phi'(a_k + 0)$. Let also for the definiteness $\phi(r) > 0$ as $r \in (a_{k-1}, a_k)$. Fix $\psi(r) \in C_0^\infty(a_{k-1}, a_{k+1})$ such that $\psi(a_k) > 0$. Denote $w(s, r) = \phi(r) + s\psi(r)$. Then, by the implicit function theorem for any sufficiently small $s : |s| < s_0 \quad w$ as a function of $r$ has a unique root $r = c(s)$ in a sufficiently small neighborhood of $a_k$; the function $c(s)$ belongs to $C^1([0, s_0))$ and $C^1((-s_0, 0])$, where we mean that at the point $s = 0$ the function $c(s)$ has derivatives on the right and on the left which are equal to $\lim_{s \to +0} c'(s)$ and $\lim_{s \to -0} c'(s)$, respectively. Let $\overline{w}(\alpha, \beta, s, r) = \alpha w(s, r)$ for $r \in [a_{k-1}, c(s))$ and $\overline{w}(\alpha, \beta, s, r) = \beta w(s, r)$ for $r \in [c(s), a_{k+1}]$, where $\alpha, \beta > 0$ and let $\alpha(s) > 0$ and $\beta(s) > 0$ be the values of $\alpha$ and $\beta$ such that the function $\overline{w}(\alpha(s), \beta(s), s, \cdot)$ satisfies condition (II.2.8) with $a = a_{k-1}, b = c(s)$ and with $a =$

$c(s), b = a_{k+1}$, respectively.  Clearly $\overline{w}(\alpha(s), \beta(s), s, \cdot) \in C^1([0, s_1); H_0^1(a_{k-1}, a_{k+1}))$
and $\overline{w}(\alpha(s), \beta(s), s, \cdot) \in C^1((-s_1, 0]; H_0^1(a_{k-1}, a_{k+1}))$ for some $s_1 > 0$.

Since by definition

$$\frac{d}{d\alpha} \int_{a_{k-1}}^{c(s)} r^{N-1} \left[ \frac{1}{2}\left(\overline{w}'^2 + \omega\overline{w}^2\right) - \frac{1}{p+1}|\overline{w}|^{p+1}\right] dr \bigg|_{\alpha=\alpha(s)} =$$

$$= \frac{d}{d\beta} \int_{c(s)}^{a_{k+1}} r^{N-1} \left[ \frac{1}{2}\left(\overline{w}'^2 + \omega\overline{w}^2\right) - \frac{1}{p+1}|\overline{w}|^{p+1}\right] dr \bigg|_{\beta=\beta(s)} = 0,$$

we easily derive using an integration by parts

$$\frac{d}{ds} \int_{a_{k-1}}^{a_{k+1}} r^{N-1} \left[ \frac{1}{2}(\overline{w}'^2 + \omega\overline{w}^2) - \frac{1}{p+1}|\overline{w}|^{p+1}\right] dr \bigg|_{s=+0} =$$

$$= \frac{d}{ds} \int_{a_{k-1}}^{a_{k+1}} r^{N-1} \left[ \frac{1}{2}(\overline{w}'^2 + \omega\overline{w}^2) - \frac{1}{p+1}|\overline{w}|^{p+1}\right] dr \bigg|_{s=-0} =$$

$$= \int_{a_{k-1}}^{a_{k+1}} r^{N-1}[\phi'\psi' + \omega\phi\psi - |\phi|^{p-1}\phi\psi]dr = a_k^{N-1}\psi(a_k)[\phi'(a_k - 0) - \phi'(a_k + 0)].$$

Hence, if $\phi'(a_k - 0) \neq \phi'(a_k + 0)$, then there exists a sufficiently small $s$ (either positive
or negative) such that

$$\int_{a_{k-1}}^{a_{k+1}} r^{N-1} \left[ \frac{1}{2}(\overline{w}'^2 + \omega\overline{w}^2) - \frac{1}{p+1}|\overline{w}|^{p+1}\right] dr <$$

$$< \int_{a_{k-1}}^{a_{k+1}} r^{N-1} \left[ \frac{1}{2}(\phi'^2 + \omega\phi^2) - \frac{1}{p+1}|\phi|^{p+1}\right] dr$$

which is a contradiction.  Thus, $\phi'(a_k - 0) = \phi'(a_k + 0)$, and Theorem II.2.1 is proved.□

## 2.3 The concentration-compactness method of P.L. Lions

The concentration-compactness method proposed by P.L. Lions (see [57,58]) has a lot
of applications to various equations (differential, integro-differential, integral, etc.) in
unbounded domains; in particular, in many cases it allows to prove the solvability of
a problem. Here we only touch upon this subject and illustrate the method with an

example of one of the simplest problems it is applicable to; we shall use the result presented below to investigate the stability of solitary waves. We consider the problem of the minimization of the functional

$$E(u) = \int_{R^N} \left\{ \frac{1}{2} |\nabla u|^2 - \frac{1}{p+1} |u|^{p+1} \right\} dx$$

under the restriction $|u|_2^2 = \lambda$ where $\lambda > 0$ is fixed; we set

$$I_\lambda = \operatorname{Inf} \left\{ E(u)/ \ u \in H^1, \ |u|_2^2 = \lambda \right\}. \tag{II.3.1}$$

The result we want to prove is the following.

<u>Theorem II.3.1</u> *Let $p \in \left(1, 1 + \frac{4}{N}\right)$ and $\lambda > 0$ be arbitrary. Then $I_\lambda > -\infty$ and for an arbitrary minimizing sequence $\{u_n\}_{n=1,2,3,\ldots}$ of the problem (II.3.1) there exists a sequence $\{y_n\}_{n=1,2,3,\ldots} \subset R^N$ such that the sequence $\{u_n(\cdot + y_n)\}_{n=1,2,3,\ldots}$ is relatively compact in $H^1$ and its arbitrary limit point is a solution of the minimization problem (II.3.1).*

<u>Remark II.3.2</u> If $p > 1 + \frac{4}{N}$, then $I_\lambda = -\infty$ for any $\lambda > 0$. For $p \in \left(1, 1 + \frac{4}{N}\right)$ and any $\lambda > 0$ one has $I_\lambda < 0$. To see this, consider the function

$$u(\sigma, x) = \frac{\sqrt{\lambda}}{(2\pi\sigma^2)^{\frac{N}{4}}} e^{-\frac{|x|^2}{4\sigma^2}}.$$

We have $|u(\sigma, \cdot)|_2^2 = \lambda$ and

$$E(u(\sigma, x)) = \frac{1}{2\sigma^2} \int_{R^N} |\nabla u(1, x)|^2 dx - \sigma^{-\frac{N}{2}(p+1)+N} \int_{R^N} \frac{1}{p+1} |u(1, x)|^{p+1} dx.$$

Hence, $E(u(\sigma, \cdot)) \to -\infty$ as $\sigma \to +0$ when $p > 1 + \frac{4}{N}$ and, for $p \in \left(1, 1 + \frac{4}{N}\right)$, $E(u(\sigma, \cdot))$ becomes negative for sufficiently large $\sigma > 0$.

<u>Remark II.3.3</u> Really, P.L. Lions in [57,58] considered problems of the essentially more general kind. For example, he investigated the problem

$$-\Delta u + c(x)u = f(x, u), \quad x \in R^N,$$

$$u\big|_{|x|\to\infty} = 0,$$

where $c(x) > 0$ and $f(x, u)$ is something like $k(x)|u|^{p-1}u$ with $k(x) > 0$.

<u>Remark II.3.4</u> As P.L. Lions noted in his publications, the principal relation providing the relative compactness of any minimizing sequence up to translations as

in Theorem II.3.1 is $I_\lambda < I_\alpha + I_{\lambda-\alpha}$ for $\alpha \in (0, \lambda)$. Below we obtain this relation for our problem.

**Lemma II.3.5** *Let* $\{u_n\}_{n=1,2,3,...}$ *be a sequence bounded in* $H^1$ *and satisfying the condition*

$$|u_n|_2^2 = \lambda,$$

*where* $\lambda > 0$ *is fixed. Then, there exists a subsequence* $\{u_{n_k}\}_{k=1,2,3,...}$ *satisfying one of the following three properties:*

   *(i) (compactness) there exists* $\{y_k\}_{k=1,2,3,...} \subset R^N$ *such that for any* $\epsilon > 0$ *there exists* $R > 0$ *for which*

$$\int_{y_k+B_R(0)} u_{n_k}^2(x + y_k)dx \geq \lambda - \epsilon, \quad k = 1, 2, 3, ...$$

*(here* $B_R(0) = \{x \in R^N : |x| < R\})$;
   *(ii) (vanishing)*

$$\lim_{k\to\infty} \sup_{y\in R^N} \int_{y+B_R(0)} u_{n_k}^2(x)dx = 0$$

*for all* $R > 0$;
   *(iii) (dichotomy) there exist* $\alpha \in (0, \lambda)$ *and sequences* $\{u_k^1\}_{k=1,2,3,...}$ *and* $\{u_k^2\}_{k=1,2,3,...}$ *bounded in* $H^1$ *and satisfying the following:*

$$\left|u_{n_k} - (u_k^1 + u_k^2)\right|_q \to 0 \quad as \quad k \to \infty \quad for \quad 2 \leq q < \frac{2N}{N-2};$$

$$\lim_{k\to\infty} \int_{R^N} (u_k^1)^2 dx - \alpha = \lim_{k\to\infty} \int_{R^N} (u_k^2)^2 dx - (\lambda - \alpha) = 0;$$

$$\text{dist}\left(\text{Supp } u_k^1; \text{Supp } u_k^2\right) \to +\infty \quad as \quad k \to \infty;$$

$$\liminf_{k\to\infty} \int_{R^N} \left\{|\nabla u_{n_k}|^2 - |\nabla u_k^1|^2 - |\nabla u_k^2|^2\right\} dx \geq 0.$$

With the Proof of Lemma II.3.5 we, actually, repeat the proof of Lemma III.1 from Part I of [57]. We introduce the concentration functions of measures

$$Q_n(t) = \sup_{y\in R^N} \int_{y+B_t(0)} u_n^2(x)dx.$$

Then, $\{Q_n(t)\}_{n=1,2,3,...}$ is a sequence of nondecreasing, nonnegative, uniformly bounded functions on $R_+$ and $\lim_{t\to+\infty} Q_n(t) = \lambda$. By the classical result, there exist a subsequence $\{Q_{n_k}\}_{k=1,2,3,...}$ and a function $Q(t)$ nonnegative and nondecreasing on $R_+$ such that $\lim_{k\to\infty} Q_{n_k}(t) = Q(t)$ for any $t \geq 0$.

Let $\alpha = \lim\limits_{t\to+\infty} Q(t)$. Obviously $\alpha \in [0, \lambda]$. If $\alpha = 0$, then the vanishing (ii) takes place for the sequence $\{Q_{n_k}(t)\}_{k=1,2,3,\ldots}$. If $\alpha = \lambda$, then clearly the compactness (i) occurs.

Let us briefly prove these two claims. First, let $\alpha = 0$ and let $\epsilon > 0$ and $R > 0$ be arbitrary. Then, we have

$$Q_{n_k}(R) = \sup_{y \in R^N} \int_{y+B_R(0)} u_{n_k}^2(x)dx < \epsilon$$

for all sufficiently large $k > 0$, and the first claim is proved. Second, let $\alpha = \lambda$. Then, obviously there exist sequences $\epsilon_k \to +0$, $R_k \to +\infty$ as $k \to \infty$ and $\{m_k\}_{k=1,2,3,\ldots}$, where $m_k$ are positive integers, such that $Q_{n_m}(R_k) > \lambda - \epsilon_k$ for all $m \geq m_k$. Take $y_k$ such that

$$\int_{y_k+B_{R_k}(0)} u_{n_k}^2(x)dx = Q_{n_k}(R_k), \quad k = 1, 2, 3, \ldots$$

Then, taking an arbitrary $\epsilon > 0$, we have for all $k$ such that $\epsilon_k < \epsilon$ and for all $m \geq m_k$:

$$\int_{B_{R_k}(0)} u_{n_m}^2(x + y_m)dx > \lambda - \epsilon.$$

Now, to get this relation for all $m = 1, 2, 3, \ldots$, it suffices to take a sufficiently large $R > R_k$ so that the second claim is proved.

Consider the case $\alpha \in (0, \lambda)$. We have to prove that the dichotomy (iii) takes place in this case. Clearly, there exist sequences $R_k \to +\infty$ and $\epsilon_k > 0$, $\epsilon_k \to 0$ as $k \to \infty$, such that

$$|Q_{n_k}(R_k) - \alpha| \leq \epsilon_k \quad \text{and} \quad |Q_{n_k}(4R_k) - \alpha| \leq \epsilon_k, \quad k = 1, 2, 3, \ldots$$

Indeed, for any positive integer $m$ there exists a number $k_m > 0$ such that

$$|Q_{n_k}(m) - Q(m)| \leq m^{-1} \quad \text{and} \quad |Q_{n_k}(4m) - Q(4m)| \leq m^{-1}$$

for all $k > k_m$. Then, we set $R_1 = \ldots = R_{k_1} = 1$, $\epsilon_1 = \ldots = \epsilon_{k_1} = 2\lambda$ and, for each $m = 1, 2, 3, \ldots$, $R_{k_m+1} = \ldots = R_{k_{m+1}} = m$, $\epsilon_{k_m+1} = \ldots = \epsilon_{k_{m+1}} = m^{-1} + \alpha - Q(m)$; since $Q(+\infty) = \alpha$, we get required sequences.

Let $\theta, \varphi$ be cut-off (infinitely differentiable) functions: $0 \leq \theta, \varphi \leq 1$, $\theta(x) \equiv 1$, $\varphi(x) \equiv 0$ for $|x| \leq 1$ and $\theta(x) \equiv 0$, $\varphi(x) \equiv 1$ for $|x| \geq 2$. Let $\theta_\mu$ and $\varphi_\mu$ denote $\theta\left(\frac{\cdot}{\mu}\right)$ and $\varphi_\mu = \left(\frac{\cdot}{\mu}\right)$, respectively. We set $M = \sup\limits_{n} ||u_n||_1$. Then, clearly there exists $C > 0$ such that

$$\left| \int_{R^N} \left[ |\nabla(\theta_R(x+y)u_{n_k}(x))|^2 - \theta_R^2(x+y)|\nabla u_{n_k}(x)|^2 \right] dx \right| \leq CR^{-2}$$

and

$$\left| \int\limits_{R^N} \left[ |\nabla(\varphi_R(x+y)u_{n_k}(x)|^2 - \varphi_R^2(x+y)|\nabla u_{n_k}(x)|^2 \right] dx \right| \le CR^{-2}$$

for all $y \in R^N$, $k = 1, 2, 3, \ldots$ and $R \ge 1$. Let $y_k \in R^N$ be such that

$$Q_{n_k}(t)(R_k) = \int\limits_{y_k + B_{R_k}(0)} u_{n_k}^2(x) dx.$$

Then, setting $u_k^1 = \theta_{R_k}(\cdot + y_k)u_{n_k}$ we get

$$\lim_{k \to \infty} \int\limits_{R^N} (u_k^1)^2 dx - \alpha = \lim_{k \to \infty} \int\limits_{R^N} \left[ |\nabla u_k^1|^2 - \theta_{R_k}^2(x+y_k)|\nabla u_{n_k}|^2 \right] dx = 0.$$

Finally, let $\varphi_k = \varphi_{4R_k}(\cdot + y_k)$ and $u_k^2 = \varphi_k u_{n_k}$. Then, we have

$$\int\limits_{R^N} (u_{n_k} - (u_k^1 + u_k^2))^2 dx \le \int\limits_{\{x \in R^N: \ |x - y_k| \in [R_k, 4R_k]\}} u_{n_k}^2(x) dx \le Q_{n_k}(4R_k) - Q_{n_k}(R_k) \to 0$$

as $k \to \infty$, therefore by the Sobolev inequality

$$|u|_q^q \le C'(|u|_2^q + |u|_2^{q+N-\frac{Nq}{2}}|\nabla u|_2^{\frac{Nq}{2}-N}), \quad 2 \le q < \frac{2N}{N-2} \ (q \ge 2 \text{ for } N = 1,2), \quad (\text{II.3.2})$$

we obtain

$$|u_{n_k} - (u_k^1 + u_k^2)|_q \to 0 \text{ as } k \to \infty, \quad 2 \le q < \frac{2N}{N-2},$$

and we get the dichotomy (iii).$\square$

Now we turn to proving Theorem II.3.1. Let $\{u_n\}_{n=1,2,3,\ldots}$ be a minimizing sequence for the problem (II.3.1). At first, we prove its boundedness in $H^1$. Indeed, the sequence $\{E(u_n)\}_{n=1,2,3,\ldots}$ is bounded. Further, we use the Sobolev embedding (II.3.2) with $q = p+1$. In this inequality, the exponent $\frac{N}{2}(p+1) - N$ belongs to $(0, 2)$ for any $p \in \left(1, 1 + \frac{4}{N}\right)$. Hence, the sequence $\{|\nabla u_n|_2\}_{n=1,2,3,\ldots}$ is bounded, therefore the sequence $\{u_n\}_{n=1,2,3,\ldots}$ is bounded in $H^1$.

Lemma II.3.6 $I_\lambda < I_\alpha + I_{\lambda-\alpha}$ for any $\lambda > 0$ and $\alpha \in (0, \lambda)$.
Proof. Let $\alpha \in [\frac{\lambda}{2}, \lambda)$ and $\theta \in (1, \frac{\lambda}{\alpha}]$. Then,

$$I_{\theta\alpha} = \inf_{u \in H^1: \ |u|_2^2 = \theta\alpha} E(u) = \inf_{u \in H^1: \ |u|_2^2 = \alpha} E(\theta^{\frac{1}{2}}u) =$$

$$= \theta \inf_{u \in H^1: \ |u|_2^2 = \alpha} \left\{ E(u) - \frac{\theta^{\frac{p-1}{2}}}{p+1} \int\limits_{R^N} |u|^{p+1} dx \right\} < \theta I(\alpha),$$

since as it is shown in Remark II.3.2 $I_\alpha = \inf\limits_{u \in H^1:\ |u|_2^2 = \alpha} E(u) < 0$. Hence,

$$I(\lambda) < \frac{\lambda}{\alpha} I(\alpha) = I(\alpha) + \frac{\lambda - \alpha}{\alpha} I(\alpha) \leq I(\alpha) + I(\lambda - \alpha).\ \square$$

Consider a subsequence $\{u_{n_k}\}_{k=1,2,3,\ldots}$ given by Lemma II.3.5. We shall show that it has the property (i) (compactness). Let us show that (iii) (dichotomy) cannot occur. Suppose the opposite. Let $\alpha_k > 0$ and $\beta_k > 0$ be such that $|\alpha_k u_k^1|_2^2 = \alpha$ and $|\beta_k u_k^2|_2^2 = \lambda - \alpha$. Then $\lim\limits_{k \to \infty} \alpha_k = \lim\limits_{k \to \infty} \beta_k = 1$ and we have $E(u_{n_k}) \geq E(u_k^1) + E(u_k^2) + \gamma_k = E(\alpha_k u_k^1) + E(\beta_k u_k^2) + \gamma_k'$ where $\gamma_k, \gamma_k' \to 0$ as $k \to \infty$. Hence,

$$I_\lambda = \lim\limits_{k \to \infty} E(u_{n_k}) \geq \lim\limits_{k \to \infty} [E(\alpha_k u_k^1) + E(\beta_k u_k^2)] \geq I_\alpha + I_{\lambda - \alpha}$$

which contradicts Lemma II.3.6.

Let us show that the vanishing (ii) cannot occur. For this aim, it suffices to prove that in the case (ii) $|u_{n_k}|_{p+1}^{p+1} \to 0$ as $k \to \infty$ because then $\liminf\limits_{k \to \infty} E(u_{n_k}) \geq 0$ which contradicts Remark II.3.2 according to which $I_\lambda < 0$. Take an arbitrary $R > 0$. Then, according to the Sobolev inequality, for any $y \in R^N$

$$|u|_{L_{p+1}(y + B_R(0))}^{p+1} \leq C(R) \left( |u|_{L_2(y + B_R(0))}^{p+1} + |u|_{L_2(y + B_R(0))}^{p+1+N-\frac{N}{2}(p+1)} \times |\nabla u|_{L_2(y + B_R(0))}^{\frac{N}{2}(p+1)-N} \right).$$

Let a sequence $\{z_r\}_{r=1,2,3,\ldots} \subset R^N$ be such that $R^N \subset \bigcup\limits_{r=1}^{\infty} \{z_r + B_R(0)\}$ and each point $x \in R^N$ is contained in at most $l$ balls where $l > 0$ is a fixed integer. Then, applying the above Sobolev inequality, we get ($\epsilon_k = \sup\limits_r |u_{n_k}|_{L_2(z_r + B_R(0))}^{p-1} \to 0$ as $k \to \infty$):

$$|u_{n_k}|_{p+1}^{p+1} \leq \sum_{r=1}^{\infty} |u_{n_k}|_{L_{p+1}(z_r + B_R(0))}^{p+1} \leq$$

$$\leq C(R)\epsilon_k \sum_{r=1}^{\infty} \left\{ |u_{n_k}|_{L_2(z_r + B_R(0))}^2 + |u_{n_k}|_{L_2(z_r + B_R(0))}^{2+N-\frac{N}{2}(p+1)} \times |\nabla u_{n_k}|_{L_2(z_r + B_R(0))}^{\frac{N}{2}(p+1)-N} \right\} \leq$$

$$\leq C_1 \epsilon_k \sum_{r=1}^{\infty} \int_{z_r + B_R(0)} [u_{n_k}^2 + |\nabla u_{n_k}|^2] dx \leq C_1 l \epsilon_k \|u_{n_k}\|_1^2 \to 0 \quad \text{as} \quad k \to \infty,$$

so that indeed the vanishing (ii) is impossible.

By the above arguments, there exists a subsequence $\{u_{n_k'}\}_{k=1,2,3,\ldots}$ of the sequence $\{u_{n_k}\}_{k=1,2,3,\ldots}$ and a sequence $\{y_k\}_{k=1,2,3,\ldots} \subset R^N$ such that the sequence $\{u_{n_k'}(\cdot + y_k)\}_{k=1,2,3,\ldots}$ converges to some $u_0 \in H^1$ strongly in $L_2$ and weakly in $H^1$. By inequality (II.3.2) this subsequence also converges to $u_0$ strongly in $L_{p+1}$. Hence, $|u_0|_2^2 = \lambda$ and $\liminf\limits_{k \to \infty} E(u_{n_k'}) \geq E(u_0)$, but then $E(u_0) = \lim\limits_{k \to \infty} E(u_{n_k'})$, hence, $\lim\limits_{k \to \infty} \|u_{n_k'}(\cdot + y_k)\|_1 = \|u_0\|_1$, therefore the sequence $\{u_{n_k'}(\cdot + y_k)\}_{k=1,2,3,\ldots}$ converges to $u_0$ strongly in $H^1$ as $k \to \infty$. This easily yields the statement of Theorem II.3.1.$\square$

## 2.4    On basis properties of systems of solutions

In this section, we briefly (with an example) consider recent results obtained by the
author which state that in certain cases systems of eigenfunctions of one-dimensional
nonlinear eigenvalue problems on a segment similar to (II.0.1),(II.0.3) can be bases in
suitable spaces of functions as $L_2$ containing "arbitrary functions". More precisely,
we consider the following nonlinear eigenvalue problem:

$$-u'' + f(u^2)u = \lambda u, \quad x \in (0,1), \quad u = u(x), \tag{II.4.1}$$

$$u(0) = u(1) = 0, \tag{II.4.2}$$

$$\int_0^1 u^2(x)dx = 1, \tag{II.4.3}$$

where again all quantities are real, $f$ is a given sufficiently smooth function, and $\lambda$ is
a spectral parameter. If a pair $(\lambda, u)$, where $\lambda \in R$ and $u = u(x)$ is a function twice
continuously differentiable on $[0,1]$, satisfies the problem (II.4.1)-(II.4.3), then we call
$\lambda$ the *eigenvalue* and $u(x)$ the corresponding *eigenfunction* of this problem. We note
that it is not clear a priori that only one eigenfunction (up to the coefficient $\pm 1$) can
correspond to an eigenvalue.

At first, we introduce some definitions which are partially known. Let $H$ be a
real separable Hilbert space with a scalar product $(\cdot, \cdot)$ and the corresponding norm
$\| \cdot \| = (\cdot, \cdot)^{\frac{1}{2}}$.

<u>Definition II.4.1.</u> *A system* $\{h_n\}_{n=0,1,2,...} \subset H$ *is called a basis of the space* $H$ *if
for an arbitrary* $h \in H$ *there exists a unique sequence of real coefficients* $\{a_n\}_{n=0,1,2,...}$
*such that* $\sum\limits_{n=0}^{\infty} a_n h_n = h$ *in the sense of the space* $H$.

<u>Definition II.4.2.</u> *A system* $\{h_n\}_{n=0,1,2,...} \in H$ *is called linearly independent if
the equality* $\sum\limits_{n=0}^{\infty} a_n h_n = 0$, *where* $a_n$ *are real coefficients, takes place in* $H$ *only for*
$0 = a_0 = a_1 = ... = a_n = ...$.

In accordance with the papers [5,6] we introduce the following two definitions.

<u>Definition II.4.3.</u> *We call a basis* $\{h_n\}_{n=0,1,2,...}$ *of the space* $H$ *a Riesz basis of
this space if the series* $\sum\limits_{n=0}^{\infty} a_n h_n$ *with real coefficients* $a_n$ *converges in* $H$ *when and only
when* $\sum\limits_{n=0}^{\infty} a_n^2 < \infty$.

<u>Definition II.4.4.</u> *Two systems* $\{e_n\}_{n=0,1,2,...} \subset H$ *and* $\{h_n\}_{n=0,1,2,...} \subset H$ *are called quadratically close in* $H$ *if* $\sum\limits_{n=0}^{\infty} ||h_n - e_n||^2 < \infty.$

Our proof of the result of this section is directly based on the following theorem of N.K. Bary.

<u>Bary Theorem.</u> *Let* $\{e_n\}_{n=0,1,2,...} \subset H$ *be a Riesz basis of* $H$ *and a system* $\{h_n\}_{n=0,1,2,...} \subset H$ *be linearly independent and quadratically close to the system* $\{e_n\}_{n=0,1,2,...}$ *in* $H$. *Then the system* $\{h_n\}_{n=0,1,2,...}$ *is a Riesz basis in* $H$.

Clearly, any orthonormal basis of the space $H$ is a Riesz basis. Bases quadratically close to orthonormal are also called in the literature *Bary bases*. Now we can establish our result on the property of being a basis for the system of eigenfunctions of the problem (II.4.1)-(II.4.3).

<u>Theorem II.4.5.</u> *Let* $f(u^2)u$ *be a continuously differentiable function of* $u \in R$ *and let* $f(r)$ *be a continuous nondecreasing function of* $r \in [0, \infty)$. *Then*
  *(a) for any integer* $n \geq 0$ *the problem (II.4.1)-(II.4.3) has a pair* $(\lambda_n, u_n)$, *consisting of an eigenvalue* $\lambda_n$ *and a corresponding eigenfunction* $u_n$, *such that the eigenfunction* $u_n$ *possesses precisely* $n$ *roots in the interval* $(0,1)$, *and this pair is unique up to the coefficient* $\pm 1$ *of the function* $u_n$; *in addition,* $\lambda_0 < \lambda_1 < ... < \lambda_n < ...$;
  *(b) the system of eigenfunctions* $\{u_n\}_{n=0,1,2,...}$ *is a Bary basis of the (real) space* $L_2(0,1)$.

Before proving this result, we shall prove the Bary theorem. For this aim, we need also a theorem of I.M. Gelfand proved in [31]. Let $X$ be a real Banach space with a norm $|| \cdot ||$. We call a real-valued functional $p$ defined on $X$ *admissible* if the following conditions are satisfied:
  1. $p(x) \geq 0$ for any $x \in X$;
  2. $p(x + y) \leq p(x) + p(y)$;
  3. $p(\alpha x) = |\alpha| p(x)$, $\alpha \in R$.
A functional $p(\cdot)$ on $X$ is called *lower semicontinuous* at a point $x_0 \in X$ if $\liminf\limits_{x \to x_0} p(x) \geq p(x_0)$.

<u>Gelfand theorem.</u> *Let* $\{p_n\}_{n=1,2,3,...}$ *be a sequence of admissible continuous functionals defined in* $X$. *If*

$$p(x) = \sup_n p_n(x) < +\infty$$

*for any* $x \in X$, *then the functional* $p(x)$ *is admissible and continuous in* $X$.

Now we turn to proving the above theorems. We begin with the <u>Proof</u> of the Gelfand theorem. It is based on the following result.

<u>Lemma II.4.6</u> *Let p be an admissible functional in X lower semicontinuous at each point of X. Then, there exists M > 0 such that*

$$p(x) \leq M||x|| \qquad\qquad (II.4.4)$$

*for all $x \in X$.*

<u>Proof of Lemma II.2.6</u>. We first prove that if the functional $p$ is bounded at least in one ball $B_r(a)$, $r > 0$, $a \in X$, then (II.4.4) holds. Indeed, if $\sup\limits_{x \in B_r(a)} p(x) < \infty$ for a ball $B_r(a)$, then

$$p(x - a) \leq p(x) + p(a) \leq C$$

for all $x \in B_r(a)$. Therefore, for any $y \in B_1(0)$, setting $y = r^{-1}(x - a)$, $r > 0$, we get

$$p(y) = r^{-1}p(x - a) \leq Cr^{-1},$$

i. e. $p$ is bounded on $B_1(0)$ and hence, due to the properties of an admissible functional, (II.4.4) holds.

Let now (II.4.4) is invalid. Then, there exists $x_1 \in B_1(0)$ such that $p(x_1) > 1$. By the lower semicontinuity of the functional $p$ there exists a ball $B_{r_1}(x_1)$ with $\overline{B_{r_1}(x_1)} \subset B_1(0)$ such that $p(x) > 1$ for all $x \in B_{r_1}(x_1)$. Then, by analogy there exists $x_2 \in B_{r_1}(x_1)$ such that $p(x_2) > 2$ and, again by the lower semicontinuity, there is a ball $B_{r_2}(x_2) : \overline{B_{r_2}(x_2)} \subset B_{r_1}(x_1)$ such that $p(x) > 2$ for any $x \in B_{r_2}(x_2)$. Continue this process. We get a sequence $\{B_{r_n}(x_n)\}_{n=1,2,3,...}$ of balls such that $\overline{B_{r_{n+1}}(x_{n+1})} \subset B_{r_n}(x_n)$ and $p(x) \geq n$ for any $x \in B_{r_n}(x_n)$; in addition, we can accept that $r_n \to 0$ as $n \to \infty$. Then, there is a unique $x_0 \in \bigcap\limits_{n \geq 1} \overline{B_{r_n}(x_n)}$. By the construction $p(x_0) \geq n$ for any integer $n > 0$. This contradiction proves the lemma.$\square$

It easily follows from Lemma II.4.6 that an admissible lower semicontinuous functional in $X$ is continuous. Indeed, let $x_0 \in X$. Then, since $p(x) \leq p(x-x_0)+p(x_0)$, we have

$$p(x) - p(x_0) \leq p(x - x_0) \leq M||x - x_0||.$$

On the other hand, by the lower semicontinuity for any $\epsilon > 0$ there exists $\delta > 0$ such that

$$p(x_0) - p(x) < \epsilon$$

for all $x : ||x - x_0|| < \delta$. Hence, for $x : ||x - x_0|| < \min\{\delta; \epsilon M^{-1}\}$ we have

$$|p(x) - p(x_0)| < \epsilon,$$

and the continuity of $p$ is proved.

Now we shall prove the Gelfand theorem. It is clear that the functional $p$ from this theorem is admissible. In view of the above arguments it suffices to show the lower semicontinuity of $p$. Let $x_0 \in X$ and $\epsilon > 0$ be arbitrary. We take a number $N \geq 0$ such that

$$p(x_0) - p_N(x_0) < \frac{\epsilon}{2}.$$

Choose $\delta > 0$ such that $p_N(x_0) - p_N(x)| < \frac{\epsilon}{2}$ for $||x - x_0|| < \delta$ (this is possible because of the continuity of $p_N$). Then, for any $x : ||x - x_0|| < \delta$ we have

$$p(x_0) - p(x) < p_N(x_0) + \frac{\epsilon}{2} - \sup_n p_n(x) \leq p_N(x_0) + \frac{\epsilon}{2} - p_N(x) < \epsilon.\square$$

<u>Corollary II.4.7</u> *Let $\{f_n(x)\}_{n=1,2,3,...}$ be a sequence of real-valued continuous linear functionals in $X$. If $\sum_{n=1}^{\infty} |f(x)|^p < \infty$ $(p \geq 1)$ for all $x \in X$, then there exists $M > 0$ such that*

$$\sum_{n=1}^{\infty} |f(x)|^p \leq M^p ||x||^p$$

*for all $x \in X$.*

<u>Proof.</u> Let $p_N(x) = \left\{ \sum_{n=1}^{N} |f_n(x)|^p \right\}^{\frac{1}{p}}$ $(N = 1, 2, 3, ...)$ and $p(x) = \sup_n p_n(x)$. By the Gelfand theorem the functional $p$ is admissible and continuous. Thus by Lemma II.4.6 $p(x) \leq M||x||.\square$

Now we present a proof of the Bary theorem. In fact, we repeat the proof from [6]. Let $h \in H$. We have

$$h = \sum_{n=0}^{\infty} a_n e_n, \quad \sum_{n=0}^{\infty} a_n^2 < \infty.$$

Let us prove that $\sum_{n=0}^{\infty} a_n[e_n - h_n]$ converges in $H$. Indeed,

$$\left\| \sum_{n=m+1}^{m+p} a_n(e_n - h_n) \right\|^2 \leq \sum_{n=m+1}^{m+p} a_n^2 \sum_{n=m+1}^{m+p} ||e_n - h_n||^2 \to 0$$

as $m \to \infty$ uniformly with respect to $p > 0$ by the convergence of the series $\sum_{n=0}^{\infty} ||e_n - h_n||^2$. Clearly, coefficients $a_n$ are linear functionals in $H$. Let us also show their continuity. We have only to prove that if $\{h^l\}_{l=1,2,3,...} \subset H$ and $\lim_{l \to \infty} ||h^l|| = 0$, then for any fixed $n$ the corresponding coefficient $a_n^l \to 0$ as $l \to \infty$. Suppose this is not the case. Then, we can assume, passing to a subsequence if necessary, that $a_n^l \geq c_0 > 0$.

Then, $g^l = (a_n^l)^{-1}h^l = \sum_{k \neq n} b_k^l e_k + e_n \to 0$ in $H$ as $l \to \infty$. But then $\sum_{k \neq n} b_k^l e_k \to g'$ in $H$ as $l \to \infty$ and clearly $g' = \sum_{k \neq n} b_k e_k$ for some real coefficients $b_k$, hence

$$e_n + \sum_{k \neq n} b_k e_k = 0 \quad \text{in} \quad H,$$

i. e. we get a contradiction. Thus indeed coefficients $a_n$ are continuous linear functionals in $H$.

Let $F = \sum_{n=0}^{\infty} a_n(e_n - h_n)$ and $F = Uf$. The operator $U$ is linear and it is determined everywhere in $H$. Let us prove that it is completely continuous. Let $B_R(0) = \{f \in H : \|f\| < R\}$ and $f \in B_R(0)$. First, we remark that according to Corollary II.4.7 there exists $M > 0$ such that $\sum_{n=0}^{\infty} a_n^2 \leq M^2\|f\|^2$ for all $f \in H$. Then, take an arbitrary $\epsilon > 0$. There exists a number $N > 0$ so large that $\sum_{n=N+1}^{\infty} \|e_n - h_n\|^2 < \epsilon$, therefore

$$\left\| F - \sum_{n=1}^{N} a_n(e_n - h_n) \right\|^2 = \left\| \sum_{n=N+1}^{\infty} a_n(e_n - h_n) \right\|^2 \leq \sum_{n=N+1}^{\infty} a_n^2 \sum_{n=N+1}^{\infty} \|e_n - h_n\|^2 \leq$$

$$\leq \epsilon \sum_{n=N+1}^{\infty} a_n^2 \leq \epsilon M^2\|f\|^2 \leq \epsilon M^2 R^2.$$

In view of the arbitrariness of $\epsilon > 0$ for any $f \in B_R(0)$ the element $F = Uf$ can be approximated arbitrary closely by functions from the compact family of functions of the kind $\sum_{n=0}^{N} a_n(e_n - h_n)$, i. e. $U(B_R(0))$ is a relatively compact subset of the space $H$ and the complete continuity of the operator $U$ is proved.

Let $\phi = f - F$, then $\phi = \sum_{n=0}^{\infty} a_n h_n$. Set $A = E - U$ where $E$ is the identity. Then $A$ is a bounded linear operator in $H$ and $\phi = Af$. But the equation

$$Af = 0 \tag{II.4.5}$$

can have only the trivial solution $f = 0$ because the system $\{h_n\}_{n=0,1,2,\ldots}$ is linearly independent. Then, since the linear homogeneous problem (II.4.5) has only the trivial solution and the operator $U$ is completely continuous, the operator $A$ has a bounded inverse $A^{-1}$ in $H$; in addition, obviously $Ae_n = h_n$.

Let $u \in H$ and $v = A^{-1}u = \sum_{n=0}^{\infty} a_n e_n$ where $\sum_{n=0}^{\infty} a_n^2 < \infty$. Then, $u = Av = \sum_{n=0}^{\infty} a_n Ae_n = \sum_{n=0}^{\infty} a_n h_n$ in $H$. Conversely, if $u = \sum_{n=0}^{\infty} a_n h_n$, then $A^{-1}u = \sum_{n=0}^{\infty} a_n e_n$, hence $\sum_{n=0}^{\infty} a_n^2 < \infty$.$\quad\square$

Remark II.4.8 As it is indicated in the proof of the Bary theorem, for any Riesz basis $\{h_n\}_{n=0,1,2,\dots}$ in $H$ there exists $m > 0$ such that

$$\|f\|^2 = \left\|\sum_{n=0}^{\infty} a_n h_n\right\|^2 \geq m^2 \sum_{n=0}^{\infty} a_n^2$$

for all $f = \sum_{n=0}^{\infty} a_n h_n \in H$. Therefore, we have a bounded one-to-one operator from $H$ onto $l_2$ which maps $f = \sum_{n=0}^{\infty} a_n h_n \in H$ into the sequence of coefficients $a = (a_0, a_1, \dots, a_n, \dots) \in l_2$. By the Banach theorem the inverse map is also bounded, i. e. there exists $M > 0$ such that

$$\left\|\sum_{n=0}^{\infty} a_n h_n\right\|^2 \leq M^2 \sum_{n=0}^{\infty} a_n^2.$$

So, we get the estimate

$$m^2 \sum_{n=0}^{\infty} a_n^2 \leq \left\|\sum_{n=0}^{\infty} a_n h_n\right\|^2 \leq M^2 \sum_{n=0}^{\infty} a_n^2$$

with positive constants $0 < m < M$ independent of $a = (a_0, a_1, \dots, a_n, \dots) \in l_2$. Another property of a Riesz (or Bary) basis is that it keeps the property to be a Riesz (resp. Bary) basis after an arbitrary reindexing of its elements (on this subject, see [39]).

Now we shall prove Theorem II.4.5. In spite of the fact that equation (II.4.1) can be solved by quadratures, we use methods of the qualitative theory of ODEs for proving the statement (a). In this part of the proof we follow our paper [114]. Without the loss of the generality we assume that $f(0) = 0$. Consider the following Cauchy problem

$$-u'' + f(u^2)u = \lambda u, \quad x \in (0,1), \tag{II.4.6}$$

$$u(0) = 0, \quad u'(0) = a > 0, \tag{II.4.7}$$

where $a > 0$ is a parameter. Denote by $u(a, \lambda, x)$ its solutions.

Lemma II.4.9 For any integer $n \geq 0$ and $a > 0$ there exists a unique $\lambda \in R$ such that the solution of the problem (II.4.6),(II.4.7) regarded as a function of the argument $x$ has precisely $n$ roots in $(0,1)$ and $u(a, \lambda, 1) = 0$; in addition $\lambda \geq (\pi(n+1))^2$.

Proof. For solutions of the problem (II.4.6),(II.4.7) the following identity takes place:

$$\{[u_x']^2 + \lambda u^2 - F(u^2)\}' = 0,$$

where $F(s) = \int\limits_0^s f(p)dp$, therefore

$$[u_x'(a, \lambda, x)]^2 + \lambda u^2(a, \lambda, x) - F(u^2(a, \lambda, x)) = a^2 \qquad (\text{II.4.8})$$

for an arbitrary solution of the problem (II.4.6),(II.4.7) and for any $x \in [0, 1]$.

Let us fix an arbitrary integer $n \geq 0$ and $a > 0$. Then, it follows from the comparison theorem that $u(a, \lambda, x) > 0$ for all $x \in (0, 1]$ and sufficiently small positive $\lambda$. At the same time, for any $A > 0$ there exists $\lambda > 0$ so large that the function $G(\lambda, u) = \lambda u^2 - F(u^2)$ strictly increases for $u \in [0, A]$ and

$$G(\lambda, A) > a^2 = [u_x'(a, \lambda, 0)]^2 + \lambda[u(a, \lambda, 0)]^2 - F(u^2(a, \lambda, 0)).$$

If a $\lambda > 0$ satisfies these conditions, then by (II.4.8) $|u(a, \lambda, x)| < A$ for all $x \in [0, 1]$. In particular, for these values of $\lambda$ the solution $u(a, \lambda, x)$ as a function of $x$ can be continued onto the whole segment $[0, 1]$ as in Section 2.1. Therefore, $\max\limits_{x \in [0,1]} |u(a, \lambda, x)| \to 0$ for $\lambda \to +\infty$. In particular, this implies that the function $u(a, \lambda, x)$ satisfies the equation

$$u'' + c(\lambda, x)u = 0, \quad x \in (0, 1),$$

where the function $c(\lambda, x) > 0$ is arbitrary large uniformly with respect to $x \in [0, 1]$ if $\lambda > 0$ is sufficiently large. Hence, by the comparison theorem the solution $u(a, \lambda, x)$ of the problem (II.4.6),(II.4.7) has more than $n$ roots in $(0, 1)$ for sufficiently large values $\lambda > 0$.

Let $\Lambda_n$ be the set of values $\lambda > 0$ such that for each of them the solution $u(a, \lambda, x)$ of the problem (II.4.6),(II.4.7) has at least $(n + 1)$ roots as a function of $x \in (0, 1)$. According to the above arguments the set $\Lambda_n$ is nonempty. Let $\lambda_n(a) = \inf \Lambda_n$. Then $\lambda_n(a) > 0$ by the above arguments. The corresponding solution $u(a, \lambda_n(a), x)$ has at most $n$ roots in $(0, 1)$ as a function of the argument $x$ because otherwise, due to the continuous dependence theorem and since $u_x'(a, \lambda, x_0) \neq 0$ if $u(a, \lambda, x_0) = 0$, there must exist values $\lambda < \lambda_n(a)$ belonging to $\Lambda_n$. By analogy $u(a, \lambda_n(a), x)$ regarded as a function of the argument $x$ has at least $n$ roots in $(0, 1)$ and $u(a, \lambda_n(a), 1) = 0$ because in the opposite case the solutions, corresponding to $\lambda \in \Lambda_n$ sufficiently close to $\lambda_n(a)$, must have at most $n$ roots in $(0, 1)$.

Let us prove the uniqueness of the above value $\lambda = \lambda_n(a)$, for which the solution $u(a, \lambda_n(a), x)$ of the problem (II.4.6),(II.4.7) has precisely $n$ roots in the interval $x \in (0, 1)$ and satisfies the condition $u(a, \lambda_n(a), 1) = 0$. Suppose that there exists $\lambda' \neq \lambda_n(a)$ satisfying these conditions. Using the autonomy of equation (II.4.6) and its invariance with respect to the changes of variables $x \to c - x$ and $u \to -u$, one can easily prove that

(a) *for any root $x_0$ of an arbitrary solution of equation* (II.4.6), *this solution is odd with respect to the point $x_0$ on an arbitrary segment* $[x_0 - b, x_0 + b] \subset [0, 1]$;

(b) *between arbitrary two nearest roots $x_1 < x_2$ of an arbitrary solution $u(x)$ of equation* (II.4.6) *there exists a unique point of extremum $x_0 = \frac{x_1+x_2}{2}$ of this solution; this solution is strictly monotone on $[x_1, x_0]$ and on $[x_0, x_2]$ and even with respect to the point $x_0$ on any segment $[x_0 - b, x_0 + b] \subset [0, 1]$ (on this subject, see also Section 2.1);*

(c) $u(a, \lambda, x) = u(a, \lambda, x + 2x_1)$ *for any $x$ such that $x, x + 2x_1 \in [0, 1]$ where $x_1$ is the minimal positive root of the function $u$.*

It follows from the properties (a)-(c) that each of two solutions $u(a, \lambda_n(a), x)$ and $u(a, \lambda', x)$ of the problem (II.4.6),(II.4.7), is monotonically increasing and achieves a maximum at the point $x = \frac{1}{2(n+1)}$. Let also for the definiteness $\lambda' > \lambda_n(a)$ and let $u_1(x) = u(a, \lambda_n(a), x)$ and $u_2(x) = u(a, \lambda', x)$. Then, in view of (II.4.8), it is clear that $u_1(x) > u_2(x)$ in a right half-neighborhood of the point $x = 0$. Let $\bar{x} > 0$ be the minimal value of the argument $x \in \left(0, \frac{1}{2(n+1)}\right)$ such that $u_1(x) = u_2(x)$ or $\bar{x} = \frac{1}{2(n+1)}$ if $u_1(x) \neq u_2(x)$ for all $x \in (0, \frac{1}{2(n+1)})$. Multiplying equation (II.4.6), written for $u_1(x)$, by $u_2(x)$, the same equation, written for $u_2(x)$, by $u_1(x)$, subtracting the obtained identities one from another and integrating the result over the segment $[0, \bar{x}]$, we get:

$$0 \geq \int_0^{\bar{x}} u_1(x)u_2(x)[f(u_1^2(x)) - f(u_2^2(x)) - \lambda_n(a) + \lambda']dx. \qquad (II.4.9)$$

But since by our supposition $u_1(x) > u_2(x)$ for $x \in (0, \bar{x})$, the right-hand side of this inequality is positive, i. e. we get a contradiction.

The property $\lambda_n(a) \geq (\pi(n + 1))^2$ follows from the comparison theorem. Thus, Lemma II.4.9 is proved. $\square$

We keep the notation $\lambda_n(a)$ for the value of the parameter $\lambda$ from Lemma II.4.9. Let $u_n(a, \lambda_n(a), x) = u_n(a, x)$. By Lemma II.4.9 these definitions are correct and $\lambda_n(a) > 0$ for any $a > 0$ and integer $n \geq 0$. Let also $\alpha_n(a) = \int_0^1 u_n^2(a, x)dx$.

<u>Lemma II.4.10</u> *For any integer $n \geq 0$ the function $\lambda_n(a)$ is nondecreasing and continuous on the half-line $a > 0$.*

Proof. Let $a_1 > a_2 > 0$. We shall prove that $\lambda_n(a_1) \geq \lambda_n(a_2)$. Suppose the contrary, i. e. that $\lambda_n(a_1) < \lambda_n(a_2)$. Let $u_1(x) = u_n(a_1, x)$ and $u_2(x) = u_n(a_2, x)$. By the properties (a)-(c) from the proof of Lemma II.4.9, each of the functions $u_1(x)$ and $u_2(x)$ increases on the segment $[0, \frac{1}{2(n+1)}]$ and $x = \frac{1}{2(n+1)}$ is the point of maximum of each of them. By (II.4.8), we have $u_1'(x_1) > u_2'(x_2)$ for any $y > 0$ for which there exist

$x_1, x_2 \in (0, \frac{1}{2(n+1)})$ satisfying $u_1(x_1) = u_2(x_2) = y$. Therefore, $u_1(x) > u_2(x)$ for all $x \in (0, \frac{1}{2(n+1)})$. Proceeding as when deriving inequality (II.4.9) and taking $\bar{x} = \frac{1}{2(n+1)}$, we get

$$0 = \int_0^{\frac{1}{2(n+1)}} u_1(x)u_2(x)[f(u_1^2(x)) - f(u_2^2(x)) - \lambda_n(a_1) + \lambda_n(a_2)]dx,$$

which is obviously a contradiction because the right-hand side here is positive. So, it is proved that the function $\lambda_n(a)$ is nondecreasing on the half-line $a > 0$.

Let us prove the continuity of the function $\lambda_n(a)$. Suppose the contrary, i. e. that there exists $a_0 > 0$ such that $\lim_{a \to a_0 + 0} \lambda_n(a) > \lambda_n(a_0)$ or $\lim_{a \to a_0 - 0} \lambda_n(a) < \lambda_n(a_0)$. Let for the definiteness the first inequality take place (the second case can be considered by analogy). Then, as one can easily verify, it follows from (II.4.6) and (II.4.8) that for each $a > a_0$ sufficiently close to $a_0$ there exists $d(a) > 0$ such that

1) $d(a) \to +0$ as $a \to a_0 + 0$;

2) $u_n(a, x) > u_n(a_0, x)$ for $x \in (0, d(a))$;

3) $u_n(a, d(a)) = u_n(a_0, d(a))$ and $\frac{d}{dx}u_n(a, d(a)) \le \frac{d}{dx}u_n(a_0, d(a))$.

Therefore $u_n(a, x) < u_n(a_0, x)$ in a right half-neighborhood of the point $x = d(a)$ (because $u_{n,xx}''(a, d(a)) < u_{n,xx}''(a_0, d(a))$). Then, as above, it follows from equality (II.4.8) that $u_n(a, x) < u_n(a_0, x)$ for all $a > a_0$ sufficiently close to $a_0$ and for all $x \in (d(a), \frac{1}{2(n+1)})$. Using the identity similar to (II.4.9) with the integral over the segment $[d(a), \frac{1}{2(n+1)}]$, we get a contradiction. So, the function $\lambda_n(a)$ is continuous, and Lemma II.4.10 is proved.□

<u>Lemma II.4.11</u> $\alpha_n(a)$ *is a strictly increasing continuous function on the half-line* $a > 0$, $\lim_{a \to +0} \alpha_n(a) = 0$ *and* $\lim_{a \to +\infty} \alpha_n(a) = +\infty$.

Proof. The continuity of the function $\alpha_n(a)$ follows from the continuity of $\lambda_n(a)$ (see Lemma II.4.10) and from the continuous dependence of solutions of the problem (II.4.6),(II.4.7) on the parameters $a$ и $\lambda$. Further, as it is proved earlier, $u_n(a, x) \to 0$ as $a \to +0$ uniformly in $x \in [0, 1]$ (see the proof of Lemma II.4.10), therefore $\lim_{a \to +0} \alpha_n(a) = 0$.

Let us prove that $\lim_{a \to +\infty} \alpha_n(a) = +\infty$. First of all, we observe that $u_{n,xx}''(a, x) \le 0$ for all $x \in (0, \frac{1}{2(n+1)})$. Indeed, if we suppose the contrary, then there exists $x_0 \in (0, \frac{1}{2(n+1)})$ such that $u_{n,xx}''(a, x_0) > 0$. But $u_{n,x}'(a, x_0) > 0$ as it was indicated earlier and also $f(u^2)u - \lambda_n(a)u$ is a nondecreasing function on the half-line $u \in [0, +\infty)$; in addition, by our supposition $f(u_n^2(a, x_0)) - \lambda_n(a)u_n(a, x_0) > 0$. Hence, we get that

$u'_{n,x}(a,x) > 0$ for all $x \in (x_0, 1]$, i. e. we have a contradiction. So, $u''_{n,xx}(a,x) \le 0$ for all $x \in (0, \frac{1}{2(n+1)})$.

Now, to prove that $\lim\limits_{a \to +\infty} \alpha_n(a) = +\infty$, it suffices to show that $u_n(a, \frac{1}{2(n+1)}) \to +\infty$ as $a \to +\infty$ (because, as it is proved above, $u_n(a,x)$ is a concave function of $x$ on the segment $[0, \frac{1}{n+1}]$). Suppose that for a sequence $a_k \to +\infty$ the following takes place: $u_n(a_k, \frac{1}{2(n+1)}) \le C < +\infty$. Consider separately the following two cases: A. $f(+\infty) = +\infty$ and B. $f(+\infty) < +\infty$.

A. Let $f(+\infty) = +\infty$. Then $\lambda_n(a) \to +\infty$ as $a \to +\infty$ (because otherwise we would get from (II.4.8) that $\lim\limits_{k \to \infty} u_n(a_k, \frac{1}{2(n+1)}) = +\infty$, i. e. a contradiction). Therefore, the functions $u_n(a_k, x)$ satisfy in the interval $x \in (0, \frac{1}{2(n+1)})$ the equations

$$u''_{n,xx} + g_k(x)u_n(a_k, x) = 0,$$

where $g_k \to +\infty$ as $k \to \infty$ uniformly with respect to $x \in (0, \frac{1}{2(n+1)})$. Hence, according to standard arguments based on the comparison theorem, each of the functions $u_n(a_k, x)$ for all sufficiently large numbers $k$ must have a root in $(0, \frac{1}{2(n+1)})$, which is a contradiction. So, the case A is impossible.

B. Let $f(+\infty) < +\infty$. Then, on the one hand, by the comparison theorem, we have $\lambda_n(a_k) \le f(+\infty) + (\pi(n+1))^2 < +\infty$. But on the other hand, as in the case A, $\lim\limits_{k \to \infty} \lambda_n(a_k) = +\infty$, i. e. we get a contradiction. So, it is proved that $\lim\limits_{a \to +\infty} \alpha_n(a) = +\infty$.

It remains to prove that $\alpha_n(a)$ is a monotonically increasing function on the half-line $a \in (0, +\infty)$. To prove this, in view of the properties (a)-(c) indicated in the proof of Lemma II.4.9, it suffices to prove that for any $a_1, a_2 : 0 < a_1 < a_2$ for all $x \in (0, \frac{1}{2(n+1)})$ the inequality $u_n(a_1, x) \le u_n(a_2, x)$ takes place. Suppose the opposite, i. e. that there exist $0 < a_1 < a_2$ such that $u_n(a_1, x) > u_n(a_2, x)$ for some $x \in (0, \frac{1}{2(n+1)})$. Let $x_1 > 0$ be the minimal point from this interval such that $u_n(a_1, x_1) = u_n(a_2, x_1)$. Let us show that $u_n(a_1, x) > u_n(a_2, x)$ in a right half-neighborhood of $x_1$. Suppose again the contrary. Then $u'_{n,x}(a_1, x_1) = u'_{n,x}(a_2, x_1)$. In view of equality (II.4.8) written for $x = x_1$, we have $\lambda_n(a_1) < \lambda_n(a_2)$, hence, in view of equation (II.4.6), $u''_{n,xx}(a_1, x_1) > u''_{n,xx}(a_2, x_1)$, i. e. we get a contradiction. So, it is proved that $u_n(a_1, x) > u_n(a_2, x)$ in a right half-neighborhood of the point $x_1$. We also note that it follows from these arguments that $u'_{n,x}(a_1, x_1) > u'_{n,x}(a_2, x_1)$ if $\lambda_n(a_1) = \lambda_n(a_2)$.

Let $x_2$ be the minimal value $x \in (x_1, \frac{1}{2(n+1)})$ such that $u_n(a_1, x) = u_n(a_2, x)$ or $x_2 = \frac{1}{2(n+1)}$ if there is no such a point in $(x_1, \frac{1}{2(n+1)})$. Repeating the procedure used when deriving inequality (II.4.9) with the integration over the segment $[x_1, x_2]$, we get:

$$0 \geq \int\limits_{x_1}^{x_2} u_n(a_1, x) u_n(a_2, x) [f(u_n^2(a_1, x)) - f(u_n^2(a_2, x)) - \lambda_n(a_1) + \lambda_n(a_2)] dx;$$

in addition, by the above arguments, we have the strict equality here if $\lambda_n(a_1) = \lambda_n(a_2)$. Thus we get a contradiction. So, it is proved that $\alpha_n(a)$ is a monotonically increasing function of the argument $a > 0$, and Lemma II.4.11 is proved.□

Lemma II.4.12 *For any integer $n \geq 0$ there exists a pair $(\lambda_n, u_n(x))$ satisfying the problem (II.4.1)-(II.4.3) and such that the function $u_n(x)$ has precisely $n$ roots in the interval $(0, 1)$ and this pair is unique up to the coefficient $\pm 1$ of the function $u_n$; in addition, the following inequalities take place:*

$$0 < \lambda_1 < \lambda_2 < ... < \lambda_n < ... \tag{II.4.10}$$

Proof. In view of Lemmas II.4.9-II.4.11, it suffices to prove inequality (II.4.10). Suppose that $\lambda_n \geq \lambda_{n+1}$ for some $n \geq 0$. The solution $u_{n+1}(x)$ strictly increases for $x \in I = [0, \frac{1}{2(n+2)}]$ and $u'_{n+1}(\frac{1}{2(n+2)}) = 0$. It cannot be that $u_{n+1}(x) \geq u_n(x)$ for all $x \in I$ because then $f(u_{n+1}^2(x)) - \lambda_{n+1} \geq f(u_n^2(x)) - \lambda_n$ for $x \in I$, hence by the comparison theorem in this case there should exist a point $x_0 \in I$ such that $u'_n(x_0) = 0$. Therefore, only one of the following two cases A and B can occur.

A. Let there exist points $0 \leq x_1 < x_2 \leq \frac{1}{2(n+2)}$ such that $u_{n+1}(x) > u_n(x)$ if $x \in (x_1, x_2)$, $u_{n+1}(x_1) = u_n(x_1)$, $u_{n+1}(x_2) \geq u_n(x_2)$, and also $u_{n+1}(x_2) = u_n(x_2)$ if $x_2 < \frac{1}{2(n+2)}$. Proceeding as when deriving (II.4.9) with the integral over the segment $[x_1, x_2]$, we get the inequality

$$0 \geq \int\limits_{x_1}^{x_2} u_{n+1}(x) u_n(x) [f(u_{n+1}^2(x)) - f(u_n^2(x)) - \lambda_{n+1} + \lambda_n] dx,$$

where, as in the proof of Lemma II.4.11, the strict inequality takes place if $\lambda_{n+1} = \lambda_n$. Thus, we get a contradiction, and the case A is impossible.

B. Let $u_{n+1}(x) \leq u_n(x)$ for all $x \in I$. Observe that $u_{n+1}(x) < u_n(x)$ for some $x_0 \in I$ (because otherwise we would have $u'_n(\frac{1}{2(n+2)}) = 0$). Further, $u_{n+1}(x) \leq u_n(x)$ for all $x \in [0, \frac{1}{n+1}]$ (it is obvious visually on a picture). Let us prove that then $\|u_{n+1}\|_{L_2} < 1$ if $\|u_n\|_{L_2} = 1$. For an arbitrary integer $k = 1, 2, ..., n+1$ consider the interval $I_k = (\frac{k-1}{n+1}, \frac{k}{n+1})$ (here $u_n(\frac{k-1}{n+1}) = u_n(\frac{k}{n+1}) = 0$). In this interval, the function $u_{n+1}$ obviously has precisely one root $\bar{x}$ (because $\frac{k}{n+2} > \frac{k-1}{n+1}$ for each $k = 2, , 3, ..., n+1$). Change the function $u_{n+1}$ on the segment $I_k$ by the function $v(x)$ equal to $u_{n+1}(\bar{x} - x + \frac{k-1}{n+1})$ for $x \leq \bar{x}$ and to $u_{n+1}(\bar{x} - x + \frac{k}{n+1})$ for $x > \bar{x}$. Repeat this procedure for each $k = 1, 2, ..., n + 1$. Then, which is obvious visually, in view of the properties (a)-(c) (see the proof of Lemma II.4.9), the graph of the function $|v(x)|$ lies under the graph of the function $|u_n(x)|$ and there is an interval $J \subset (0, 1)$ such that $|v(x)| < |u_n(x)|$ for all $x \in J$. Therefore,

$$|u_{n+1}|_{L_2(0,1)} = |v|_{L_2(0,1)} < |u_n|_{L_2(0,1)} = 1,$$

i. e. we get a contradiction. Thus, Lemma II.4.12 is proved. $\square$

Let $e_n(x) = \sqrt{2}\sin[\pi(n+1)x]$, $n = 0, 1, 2, ....$ Then, the system $\{e_n\}_{n=0,1,2,...}$ is an orthonormal basis in $L_2(0, 1)$. We also accept that $u'_n(0) > 0$ for all $n = 0, 1, 2, ...$ which is possible due to the invariance of equation (II.4.1) with respect to the multiplication of $u$ by $-1$. In the next part of proving Theorem II.4.5 we follow the paper [115].

**Lemma II.4.13** *There exists $C > 0$ such that $|u_n - e_n|_2 \leq C(n+1)^{-1}$ for all $n = 0, 1, 2, ...$; in particular, the systems $\{u_n\}_{n=0,1,2,...}$ and $\{e_n\}_{n=0,1,2,...}$ are quadratically close in $L_2(0,1)$.*

Proof. At first, we shall show that the functions $\{u_n(x)\}_{n=0,1,2,...}$ are bounded uniformly with respect to $x \in [0, 1]$. Consider the sequence $u_n(b_n, x)$ of solutions of the Cauchy problem (II.4.6),(II.4.7) with $b_n = 10\pi(n + 1)$. Let us prove that this sequence is uniformly bounded. Let $d_n = u_n(b_n, \frac{1}{2(n+1)})$. Then, obviously it suffices to show the boundedness of the sequence $d_n$ from above. By the comparison theorem, for each number $n$ we have $\lambda_n(b_n) \geq (\pi(n + 1))^2$. Further, identity (II.4.8) yields the inequality

$$(10\pi(n+1))^2 \geq G(\lambda_n(b_n), u_n(b_n, x)) \geq (\pi(n+1))^2 u_n^2(b_n, x) - F(u_n^2(b_n, x)), \quad \text{(II.4.11)}$$

which implies that

$$d_n < 20$$

for all sufficiently large numbers $n$. Indeed, if we suppose that this is not so, then we would get that in (II.4.11) the right-hand side is greater than the left-hand one for all sufficiently large numbers $n$ and for all $x$ satisfying $|u_n(b_n, x)| = 20$, i. e. a contradiction. So, the uniform boundedness of the sequence $\{u_n(b_n, x)\}_{n=0,1,2,...}$ is proved.

Let us show that $|u_n(b_n, \cdot)|_{L_2(0,1)} > 1$ for all sufficiently large numbers $n$. Set $h_n(x) = 10\sin[\pi(n+1)x]$ and observe that $u'_{n,x}(b_n, 0) = h'_n(0)$ and, in view of the properties (a)-(c) from the proof of Lemma II.4.9, $u'_{n,x}(b_n, 1) = h'_n(1)$. We have:

$$-h''_n = \mu_n h_n, \quad x \in (0, 1),$$

$$h_n(0) = h_n(1) = 0$$

where $\mu_n = (\pi(n+1))^2$. Therefore, we get from the previous equation and equation (II.4.6) for the functions $w_n(x) = u_n(b_n, x) - h_n(x)$:

$$-w_n'' + W_n(x) = \mu_n w_n, \quad x \in (0,1),$$

$$w_n(0) = w_n(1) = w_n'(0) = w_n'(1) = 0,$$

where the family of functions $\{W_n(x)\}$ is uniformly bounded because of the uniform boundedness of the family of functions $\{u_n(b_n, x)\}_{n \geq 0}$ and the estimate

$$|\lambda_n - \mu_n| \leq C_1 \tag{II.4.12}$$

with a constant $C_1 > 0$ independent of $n$ taking place by the comparison theorem. Multiplying this equation by $2xw_n'(x)$, integrating the obtained equality over the segment $[0,1]$ and applying the integration by parts, we get

$$\mu_n |w_n|_{L_2(0,1)}^2 \leq -|w_{n,x}'|_{L_2(0,1)}^2 - 2\int_0^1 xw_{n,x}'(x)W_n(x)dx \leq \int_0^1 x^2 W_n^2(x)dx \leq C_2,$$

where $C_2 > 0$ is independent of $n$. In view of this equality and since $|h_n|_{L_2(0,1)} > 1$ for all numbers $n$, we have $|u_n(b_n, \cdot)|_{L_2(0,1)} > 1$ for all sufficiently large numbers $n$. This fact and Lemma II.4.11 yield that

$$|u_n'(0)| < b_n \tag{II.4.13}$$

for all sufficiently large numbers $n$.

It can be proved using (II.4.13) as for the functions $u_n(b_n, \cdot)$ that the family of the functions $\{u_n(\cdot)\}_{n=0,1,2,...}$ is uniformly bounded. Let us prove that

$$|u_n - e_n|_{L_2(0,1)} < \overline{C}(n+1)^{-1}, \quad n = 0, 1, 2, ....$$

For this aim, we first shall show that there exists $C_3 > 0$ such that

$$|u_n'(0) - e_n'(0)| \leq C_3 \quad \text{and} \quad |u_n'(1) - e_n'(1)| \leq C_3 \tag{II.4.14}$$

for all numbers $n$.

At first, let us multiply equation (II.4.1) written for $u(x) = u_n(x)$ by $u_n(x)$ and integrate the obtained identity over the segment $[0,1]$. Then, due to the uniform boundedness of the family of functions $\{u_n(x)\}_{n=0,1,2,...}$, we get

$$\lambda_n - C_4 \leq \int_0^1 [u_n'(x)]^2 dx \leq \lambda_n + C_4$$

for a constant $C_4 > 0$ independent of $n$. Further, let us multiply equation (II.4.1) written for $u(x) = u_n(x)$ by $2(1+x)u'_n(x)$ and integrate the obtained identity between 0 and 1. We get after transformations with the use of the previous estimate

$$\left|2[u'_n(1)]^2 - [u'_n(0)]^2 - 2\lambda_n\right| \leq$$

$$\leq C_5 + 2\left|\int_0^1 (1+x)f(u_n^2(x))u_n(x)u'_n(x)dx\right| \leq$$

$$\leq C_5 + \int_0^1 |F(u_n^2(x))|dx,$$

where again $F(s) = \int_0^s f(t)dt$. Hence, since due to the properties (a)-(c) from the proof of Lemma II.4.9 $[u'_n(0)]^2 = [u'_n(1)]^2$, we have

$$\left|[u'_n(1)]^2 - 2\lambda_n\right| = \left|[u'_n(0)]^2 - 2\lambda_n\right| \leq C_6$$

for a constant $C_6 > 0$ independent of $n$. Now, since sign $u'_n(0) = $ sign $e'_n(0)$ and, as it follows from the properties (a)-(c) from the proof of Lemma II.4.9, sign $u'_n(1) = $ sign $e'_n(1)$, because $[e'_n(0)]^2 = [e'_n(1)]^2 = 2\mu_n$ and in view of inequality (II.4.12), we get (II.4.14).

Let $g_n(x) = u_n(x) - e_n(x)$. Then, due to the uniform boundedness of the family of functions $\{u_n(x)\}_{n=0,1,2,...}$, we have

$$-g''_n(x) + G_n(x) = \mu_n g_n(x), \quad x \in (0,1), \tag{II.4.15}$$

$$g_n(0) = g_n(1) = 0, \tag{II.4.16}$$

where $\{G_n(x)\}_{n=0,1,2,...}$ is a uniformly bounded sequence of continuous functions. Multiplying equation (II.4.15) by $2xg'_n(x)$ and integrating the obtained identity from 0 to 1 with the use of the integration by parts, due to (II.4.14) and (II.4.16) we get

$$\mu_n \int_0^1 [g_n(x)]^2 dx \leq C_7 - \int_0^1 [g'_n(x)]^2 dx - 2\int_0^1 xG_n(x)g'_n(x)dx,$$

where $C_7 > 0$ is a constant independent of $n$. Therefore, applying the inequality $2ab \leq a^2 + b^2$, we obtain that there exists $C_8 > 0$ such that $\mu_n \int_0^1 g_n^2(x)dx \leq C_8$ for all numbers $n$. Hence, $\|g_n\| \leq C_8^{\frac{1}{2}}\pi^{-1}(n+1)^{-1}$, and Lemma II.4.13 is proved. $\square$

To prove Theorem II.4.5, now it suffices to show the linear independence in $L_2(0,1)$ of the system of functions $\{u_n\}_{n=0,1,2,\ldots}$. For this aim, let us for each integer $n \geq 0$ expand the function $u_n\left(\frac{\cdot}{n+1}\right)$ in the space $L_2(0,1)$ in the Fourier series: $u_n\left(\frac{\cdot}{n+1}\right) = \sum_{k=0}^{\infty} a_k^n e_k(\cdot)$ where $a_k^n$ are real coefficients. Then

$$u_n(\cdot) = \sum_{m=0}^{\infty} b_m^n e_m(\cdot) \tag{II.4.17}$$

in the space $L_2(0,1)$, where $b_{(n+1)(k+1)-1}^n = a_k^n$ and $b_m^n = 0$ if $m \neq (n+1)(k+1)-1$ for all $k = 0,1,2,\ldots$. Indeed, equality (II.4.17) obviously takes place in the space $L_2\left(0,\frac{1}{n+1}\right)$. Then, since in view of the properties (a)-(c) from the proof of Lemma II.4.9 the function $u_n(x)$ is odd with respect to its roots which are precisely the points $\frac{r}{n+1}$ where $r = 1,2,\ldots,n$ and since the direct verification shows that the functions $e_{(n+1)(k+1)-1}(x)$, where $k = 0,1,2,\ldots$, are odd with respect to these points, too, equality (II.4.17) also holds in the sense of each of the spaces $L_2\left(\frac{r}{n+1},\frac{r+1}{n+1}\right)$ where $r = 1,,2,\ldots,n$. Therefore, it is valid in the sense of the space $L_2(0,1)$. Also, obviously $b_0^n = \ldots = b_{n-1}^n = 0$ for each $n$ and, in view of our acceptation, $b_n^n = a_0^n > 0$ because the functions $u_n$ and $e_n$ are of the same sign everywhere.

Let us suppose that the system $\{u_n\}_{n=0,1,2,\ldots}$ is not linearly independent. Then, there exist real coefficients $c_n$, $n = 0,1,2,\ldots$, not all equal to zero such that

$$\sum_{n=0}^{\infty} c_n u_n = 0 \tag{II.4.18}$$

in the space $L_2(0,1)$. Let the index $l \geq 0$ be such that $c_0 = \ldots = c_{l-1} = 0$ and $c_l \neq 0$. Multiply equality (II.4.18) by $e_l$ in the space $L_2(0,1)$. In view of (II.4.17) we get: $b_l^l c_l = 0$. But as it is proved above $b_l^l \neq 0$ and as we supposed $c_l \neq 0$. So, we get a contradiction, and the linear independence of the system of functions $\{u_n(x)\}_{n=0,1,2,\ldots}$ in the space $L_2(0,1)$ is proved. Thus, Theorem II.4.5 is proved, too.$\square$

## 2.5   Additional remarks.

There is a large number of publications devoted to the questions of the existence of radial solutions of the problem (II.0.1),(II.0.3). In the papers [71],[96], this problem with $f(\phi^2) = |\phi|^{p-1}$ ($p > 1$) is considered. In the paper [71], by using a variational method, for $N = 3$ and $1 < p < 4$ the existence of a positive solution is proved. Independently, in [96] for the same function $f$, $1 < p \leq 3$ and $N = 3$ a proof of existence of a positive solution, based on methods of the qualitative theory of ODEs, is contained. In the paper [97] a refinement of the technique of the paper [71] is presented; the existence of a solution with an arbitrary given number of roots on the half-line $r = |x| > 0$ is proved for $N = 3$ and $1 < p < 4$. A similar result is proved in [82] for nonlinearities of a more general kind. In addition, for $p = 5$ in the paper [71] and for $p \geq 5$ in the paper [97], it is proved that there are no nontrivial solutions $(N = 3)$.

In fact, variational methods made it possible for $N = 3$ and $f(\phi^2) = |\phi|^{p-1}$ to prove the existence of a radial solution of the problem (II.0.1),(II.0.3) with an arbitrary given number of roots for $1 < p < 5$, but for $4 \leq p < 5$ the proof was left incomplete: there remained the question whether the considered solutions are bounded in a neighborhood of the point $r = 0$. This problem was completely solved in the paper [83]: it has been proved in this paper that for $1 < p < 5$ any positive solution is bounded with its first derivative in a neighborhood of the point $r = 0$, i. e. it satisfies the problem (II.0.1),(II.0.3). In fact, methods of this paper also hold for solutions with the alternative sign so that we have the result that for any $1 < p < 5$ the problem (II.0.1),(II.0.3) with $N = 3$ and $f(\phi^2) = |\phi|^{p-1}$ possesses a solution with any given number of roots on the half-line $r > 0$. A result similar to this is independently obtained in the paper [59]. Methods of the latter two papers are also applicable for investigations of radial solutions for $N \neq 3$ and for nonlinearities $f(\phi^2)\phi$ similar in a sense to $f(\phi^2) = |\phi|^{p-1}\phi$. Theorem II.1.2 was first proved in [98]; here for its proving we exploited methods from [110].

In the framework of the ODE approach, the problem of the existence of solutions of (II.0.1),(II.0.3) with $N = 3$ and $f(\phi^2) = |\phi|^{p-1}$, $p \in (3,5)$, has remained unsolved for a long time. In the paper [10], an ODE approach was exploited for this aim but the principal result on the existence of solutions of the Cauchy problem (II.1.4),(II.1.5) achieving the zero value on the half-line $r > 0$ was obtained by a variational method. The indicated problem was completely solved by methods of the qualitative theory of ODEs by B.L. Shekhter, see [47].

Another way of proving the existence of positive radial solutions was proposed by W. Strauss in [87]; his method is based on the concept of a symmetrization of a nonnegative (non-radial) function from $H^1$. Unfortunately, the author has made a mistake stating that the existence of a solution with an arbitrary given number of roots follows directly from results of P.H. Rabinowitz [77,78]; it is not so.

Till now, we have reviewed, with the model example $f(\phi^2) = |\phi|^{p-1}$, the results on the existence of radial solutions of the problem (II.0.1),(II.0.3) in the case $f(\phi^2) \to +\infty$ as $|\phi| \to \infty$. Another case is considered with Theorem II.1.2. One more case is investigated in the recent paper [110]. In this paper, it is supposed that the function $g(\phi)$ has a unique positive root, $\omega - f(0) > 0$ and there exists a finite $\lim_{|\phi|\to\infty} [\omega - f(\phi^2)] \leq 0$; sufficient conditions for the existence of a radial solution with an arbitrary given number of roots on the half-line $r = |x| > 0$ are obtained by using methods of the qualitative theory of ODEs.

The methods of our proving Theorem II.2.1 are close to the methods introduced in the papers [2,3,9,71,75,76,82,83,97]. However, to the best of our knowledge there is no paper containing precisely our variant of the proof.

The second problem (not considered in this chapter) concerns the uniqueness of solutions. In the literature, there are only results on the uniqueness of positive radial solutions of the problem (II.0.1),(II.0.3). In particular, for the nonlinearity $g(\phi) = \omega\phi - |\phi|^{p-1}\phi$, the uniqueness of the positive solution is proved in the papers [25,54,68]. The final result is obtained in [54]: as it is proved (for $g(\phi) = \omega\phi - |\phi|^{k-1}\phi$, $\omega > 0$), the positive solution is always unique (if it exists). We also note that the proof of a similar result presented in [83] contains a principal error on page 111.

Concerning the result from Section 2.4 on basis properties of the system of eigenfunctions of a nonlinear Sturm-Liouville-type problem, to the author's best knowledge this is quite a new field and there are almost no results in this direction. We only mention the monograph by A.P. Makhmudov [62] containing some interesting results on this subject and the paper by the same author [63] where the completeness of the system of eigenelements of a nonlinear operator equation, arising under small nonlinear perturbations of a linear problem, is proved. The Bary theorem was in the first announced (without a proof) in [5] and proved in [6]. A more thorough discussion of questions around this theorem is contained in [39]; the approach of this monograph to this theorem differs from the approach in [6]. The first proof of Theorem II.4.5 is established in [114]. However, in that paper the proof contains essential errors which, fortunately, can be corrected. These corrections are published in [119]. Similar problems and corresponding results on the property of being a basis for their systems of eigenfunctions in $L_2$ are presented in [115-117]. In [118], a boundary-value problem (without a spectral parameter) is considered. In the paper [118] it is proved that the system of its solutions, which is denumerable, is a basis in $H^s(0,1)$ where $s < s_0$ and $s_0$ is a negative constant. Also, in [119] an analog of the Fourier transform on the half-line $x > 0$ over eigenfunctions of a nonlinear eigenvalue problem is considered.

Here we have presented a proof of Theorem II.4.5 based on the Bary theorem. Another natural way consists in an attempt to use the expansions (II.4.17) for this aim. This approach is exploited in [118]. However, we note that an example from this paper shows that in general the properties of the coefficients $b_m^n$ (the matrix $(b_m^n)_{n,m=0,1,2,\ldots}$ is upper triangular and all elements of its principal diagonal are nonzero) are insufficient even for the system of functions $\{u_n\}_{n=0,1,2,\ldots}$ to be complete in $L_2(0,1)$.

# Chapter 3

# Stability of solutions

In this chapter, we shall consider questions of the stability of solitary waves. As it is noted in the Introduction, usually solitary waves of the KdVE and NLSE are unstable with respect to distances of standard spaces of functions as Lebesgue or Sobolev spaces and it is natural to study the stability of solitary waves vanishing as $|x| \to \infty$ with respect to the distance $\rho$ in the case of the KdVE and to $d$ for the NLSE (for definitions of $\rho$ and $d$ see Introduction or Section 3.1). We also recall that we named a solitary wave $u(x,t)$, where $u(x,t) = \phi(\omega, x - \omega t)$ in the case of the KdVE and $u(x,t) = e^{i\omega t}\phi(\omega, x)$ for the NLSE, a kink if $\phi'_x(\omega, x) \neq 0$ for all $x \in R$ and a soliton-like solution if there is a unique $x_0 \in R$ such that $\phi'_x(\omega, x_0) = 0$, $x_0$ is a point of extremum of $\phi(\omega, x)$ as a function of the argument $x$ and $\phi(-\infty) = \phi(+\infty)$.

In the mathematical literature devoted to the stability of solitary waves, the pioneer paper by T.B. Benjamin [7] was the origin which initiated further numerous investigations in the field. In this paper, the author has proved the stability of soliton-like solutions vanishing as $x \to \pm\infty$ of the standard KdVE with $f(u) = u$ with respect to the distance $\rho$; he called this stability the *stability of the form* of a soliton-like solution. This terminology can be easily understood visually: if a travelling wave $\phi(x - \omega t)$ vanishing as $x \to \pm\infty$ and a solution $u(x,t)$ of the standard KdVE are close to each other in the sense of the distance $\rho$ for some $t \geq 0$, then clearly these functions can be not close to each other in the sense of distances of usual spaces of functions as Lebesgue or Sobolev spaces, however, for this $t \geq 0$ there is a translation $\tau = \tau(t) \in R$ such that the graphs of $u(x - \tau, t)$ and $\phi(x - \omega t)$ as functions of $x$ are almost identical, i. e. the "forms of the graphs" of the functions $\phi(x - \omega t)$ and $u(x,t)$ "almost coincide". Thus, if a soliton-like solution vanishing as $|x| \to \infty$ is stable, then for each $t \geq 0$ the forms of its graph and of the graph of an arbitrary perturbed solution sufficiently close to it in the sense of the distance $\rho$ "almost coincide". In Section 3.1 we consider two sufficient conditions for the stability with respect to the distance $\rho$ of soliton-like solutions vanishing as $x \to \pm\infty$ for our KdVE. Simultaneously we study the NLSE which we suppose to be one-dimensional, i. e. with $N = 1$.

As it is noted in Section 2.1, the KdVE under natural assumptions on the behavior as $x \to \pm\infty$ of derivatives in $x$ of solutions and the NLSE with $N = 1$ can have solitary waves of only two types: these are soliton-like solutions and kinks. In Section 3.2 we prove the stability of kinks of the KdVE with respect to the distance $\rho$ under assumptions of general type.

In Section 3.3 we consider a stability of solitary waves of the NLSE nonvanishing as $x \to \pm\infty$. In two cases we prove a stability of a new interesting type.

## 3.1   Stability of soliton-like solutions

In this section, the stability of soliton-like solutions vanishing as $x \to \pm\infty$ will be considered. We begin with defining the stability of such solutions for the KdVE and one-dimensional NLSE. Let

$$\rho(u,v) = \inf_{\tau \in R} \|u(\cdot) - v(\cdot - \tau)\|_1, \quad u,v \in H^1$$

and

$$d(u,v) = \inf_{\tau \in R, \ \gamma \in [0,2\pi]} \|u(\cdot) - e^{i\gamma}v(\cdot - \tau)\|_1, \quad u,v \in H^1,$$

where $H^1$ is the real space in the first case and complex in the second. One can easily prove that in each case the greatest lower bound is achieved.

First, we consider the Cauchy problem for the KdVE:

$$u_t + f(u)u_x + u_{xxx} = 0, \quad x,t \in R, \tag{III.1.1}$$

$$u(x,0) = u_0(x); \tag{III.1.2}$$

we remark that the proof of the uniqueness of a $H^2$-solution from Theorem I.1.3 holds for the problem (III.1.1),(III.1.2) with an arbitrary twice continuously differentiable $f(\cdot)$. We also present the following result here.

<u>Proposition III.1.1</u> *Let $f(\cdot)$ be a twice continuously differentiable function and let $u(\cdot,t)$ be a $H^2$-solution of the problem (III.1.1),(III.1.2) in an interval of time $I = [0,a)$ or $I = [0,a]$, $a > 0$ ($u_0 \in H^2$ if $a = 0$). If there exists $C > 0$ such that $\|u(\cdot,t)\|_1 \le C$ for all $t \in I$, then there exists $\delta > 0$ such that the solution $u(\cdot,t)$ can be continued onto the interval $[0, a + \delta)$ (resp. there exists a $H^2$-solution of the problem (III.1.1),(III.1.2) in an interval of time $[0,\delta)$, $\delta > 0$, if $a = 0$).*

<u>Proof.</u> Let $f(\cdot)$ be a twice continuously differentiable function and $u(\cdot,t)$ be a $H^2$-solution of the problem (III.1.1),(III.1.2) in an interval of time $I = [0,a)$ or $I = [0,a]$, $a > 0$, bounded in $H^1$: $\|u(\cdot,t)\|_1 \le C$ (the case $a = 0$ can be considered by analogy). Let $C_1 > 0$ be a constant, existing in view of the embedding of $H^1$ into $C$, such that

$|g(\cdot)|_C \le C_1$ for all $g \in H^1$ obeying $||g||_1 \le C$ and let $f_1(\cdot)$ be a twice continuously differentiable function, coinciding with $f(\cdot)$ for $u \in [-1 - C_1, 1 + C_1]$ and satisfying condition (I.1.3), so that the problem (III.1.1),(III.1.2) taken with $f = f_1$ has a unique global $H^2$-solution $u_1(\cdot, t)$. Then obviously, since $u(\cdot, t)$ is a $H^2$-solution of the latter problem in $I$ and due to the uniqueness of such a solution, $u(\cdot, t) = u_1(\cdot, t)$ for $t \in I$. Also, obviously there exists $\delta > 0$ such that $|u_1(\cdot, t)|_C \le 1 + C_1$ for $t \in [0, a + \delta)$. Thus $u_1(\cdot, t)$ is a $H^2$-solution of the problem (III.1.1),(III.1.2) in the interval of time $[0, a + \delta)$. $\square$

<u>Definition III.1.2</u> *Let* $f(\cdot) \in \overset{\infty}{\underset{n=1}{\bigcup}} C^2((-n, n); R)$, *so that in accordance with Proposition III.1.1 for any* $u_0 \in H^2$ *the problem (III.1.1),(III.1.2) has a unique local $H^2$-solution, and let* $\phi(\omega, x - \omega t)$, $\omega \in R$, *be a soliton-like solution of the KdVE,* $\phi(\omega, \cdot) \in H^1$. *Then we call this solution $\phi$ stable if for any $\epsilon > 0$ there exists $\delta > 0$ such that, if $u_0 \in H^2$ and $\rho(u_0(\cdot), \phi(\omega, \cdot)) < \delta$, then the corresponding solution $u(x, t)$ of the problem (III.1.1),(III.1.2) can be continued onto the entire half-line $t > 0$ and one has $\rho(u(\cdot, t), \phi(\omega, \cdot)) < \epsilon$ for all $t > 0$.*

By analogy, consider the Cauchy problem for the NLSE:

$$iu_t + \Delta u + f(|u|^2)u = 0, \quad x, t \in R, \tag{III.1.3}$$

$$u(x, 0) = u_0(x). \tag{III.1.4}$$

<u>Definition III.1.3</u> *Let $N = 1$, the nonlinearity $f(|u|^2)u$ in equation (III.1.3) satisfy condition (f1) from Section 1.2 and let $\overline{u}(x, t) = e^{i\omega t}\phi(x)$ be a soliton-like solution of the NLSE (III.1.3) where $\phi \in H^1$. Then, we call this solution $\overline{u}(x, t)$ stable if for any $\epsilon > 0$ there exists $\delta > 0$ such that for any $u_0 \in H^1$ satisfying the condition $d(u_0, \phi) < \delta$ the corresponding $H^1$-solution $u(x, t)$ of the problem (III.1.3),(III.1.4) can be continued onto the entire half-line $t > 0$ and for all $t > 0$ one has $d(u(\cdot, t), \overline{u}(\cdot, t)) < \epsilon$ (the local existence and uniqueness of this solution $u(x, t)$ are proved with Theorem I.2.4).*

Now we consider an application of the concentration-compactness method of P.L. Lions to the investigation of the stability of soliton-like solutions. We consider the KdVE with $f(u) = |u|^\nu$, $\nu > 0$. When formulating the following result on the stability we assume that $\nu \in [2, 4)$; the requirement $\nu \ge 2$ is connected with assumptions of Theorem I.1.3 and Proposition III.1.1 according to which $f(\cdot)$ is a twice continuously differentiable function. However, if one proves or supposes the local well-posedness of the problem (III.1.1),(III.1.2) in a suitable sense (for example, in the sense of the space $H^1$) for other values of the parameter $\nu \in (0, 4)$, then all the arguments from the proof of the theorem below hold.

Theorem III.1.4 *Let $f(u) = |u|^\nu$ where $\nu \in [2,4)$. Then, for any $A > 0$ the soliton-like solution $\phi(x,t)$ of the KdVE (III.1.1) from the family (II.1.3) is stable.*

Proof. Let us fix an arbitrary $A > 0$. Suppose that the corresponding solution $\phi(x,t)$ from the family (II.1.3) is not stable. Then, there exist $\epsilon > 0$, a sequence $\{t_n\}_{n=1,2,3,...} \subset R_+$ and a sequence $\{u_0^{(n)}\}_{n=1,2,3,...}$ from $H^2$ such that $\rho(u_0^{(n)}, \phi|_{t=0}) \to 0$ as $n \to \infty$ and $\rho(u_n(\cdot,t_n), \phi|_{t=t_n}) > \epsilon$ where $u_n(x,t)$ are $H^2$-solutions of the Cauchy problem (III.1.1),(III.1.2) with $u_0 = u_0^{(n)}$ given by Theorem I.1.3.

Let $|\phi(\cdot,t)|_2^2 = \lambda > 0$. Consider the minimization problem

$$I_\lambda = \inf_{u \in H^1, \ |u|_2^2 = \lambda > 0} E(u),$$

where the functional $E(u)$ is defined in Section 2.3. According to Theorem II.3.1, it has a solution and clearly every its solution satisfies the following equation with some $\omega > 0$ and boundary conditions

$$u'' + \frac{1}{\nu+1}|u|^\nu u = \omega u, \qquad u(\pm\infty) = 0$$

(the parameter $\omega$ must be positive because otherwise, as it is shown in Section 2.1, the above problem has no nontrivial solutions). As at the beginning of Section 2.1, this boundary-value problem has a unique, up to translations, positive solution $\overline{\phi}$ that belongs to the family (II.1.3) with some $A_1 > 0$. Further, since $|\phi_1|_2^2 \neq |\phi_2|_2^2$ for any two functions $\phi_1$ and $\phi_2$ from the family (II.1.3) with different values of the parameter $A$, we get that $A_1 = A$ and $\overline{\phi} = \phi$ (up to a translation). Therefore, according to Theorem II.3.1, for any minimizing sequence $\{w_n\}_{n=1,2,3,...}$ of our minimization problem there exists a sequence $\{y_n\}_{n=1,2,3,...} \subset R$ such that $w_n(\cdot + y_n) \to \phi$ in $H^1$ as $n \to \infty$.

Return to our sequence of solutions $\{u_n\}_{n=1,2,3,...}$ of the problem (III.1.1),(III.1.2). For any $n$ we take $v_n = \frac{\lambda u_n(\cdot,t_n)}{|u_n(\cdot,t_n)|_2}$. Since according to Theorem I.1.3 $|u_n(\cdot,0)|_2^2 = |u_n(\cdot,t_n)|_2^2 \to \lambda$ as $n \to \infty$ and $E(u_n(\cdot,0)) = E(u_n(\cdot,t_n))$, it is clear that $\{v_n\}_{n=1,2,3,...}$ is a minimizing sequence for the above minimization problem, therefore there exists a sequence $\{y_n\}_{n=1,2,3,...} \subset R$ such that $v_n(\cdot + y_n)$ converges in $H^1$ to $\phi$ as $n \to \infty$. Hence, $\rho(v_n, \phi) \to 0$ as $n \to \infty$. Since $\frac{\lambda}{|u_n(\cdot,t_n)|_2} \to 1$, we get $\rho(u_n(\cdot,t_n), \phi) \to 0$ as $n \to \infty$, i. e. a contradiction. Thus, Theorem III.1.4 is proved.□

Remark III.1.5 For the first time, Theorem III.1.4 was proved in the paper [101]. In fact, in the papers by T. Cazenave and P.L. Lions [23] and by P.L. Lions [57,58] the results on the stability with proofs based on the concentration-compactness method for a NLSE of the essentially more general kind are considered: in particular, a multidimensional NLSE, admitting coefficients depending on the spatial variable, is investigated in these papers. Here we wanted only to illustrate with an example the possibility of applying this method to the problem of the stability of solitary waves.

Now we consider the "$Q$- criterion" of the stability of soliton-like solutions vanishing as $|x| \to \infty$ which gives sufficient conditions of the stability close to necessary. The name "$Q$-criterion" originates from the physical literature in which the conservation law $P$ of the NLSE ($E_0$ in the case of the KdVE) used in the condition of the stability has been often denoted by $Q$. We also rename here the conservation law $E_0$ of the KdVE by $P$ and recall that, in the case on the KdVE, $\overline{f}(u) = \int_0^u f(p)dp$ and

$$F(u) = \int_0^u \overline{f}(p)dp.$$

<u>Theorem III.1.6</u> *Let $f(|u|^2)u$ be a continuously differentiable function of the complex argument $u$ for the NLSE (III.1.3) with $N = 1$ ($f(u)$ be twice continuously differentiable for the KdVE). Let also $F(r) = \int_0^r f(s)ds$ and let there exist $\omega_0 \in R$ and $b > 0$ such that $f(0) - \omega_0 < 0$, $f(b^2) - \omega_0 > 0$, $F(b^2) - \omega_0 b^2 = 0$ and $F(\phi^2) - \omega_0 \phi^2 < 0$ for $\phi \in (0,b)$ (resp., $f(0) - \omega_0 < 0$, $\overline{f}(b) - \omega_0 b > 0$, $F(b) - \frac{\omega_0}{2}b^2 = 0$ and $F(\phi^2) - \frac{\omega_0}{2}\phi^2 < 0$ for $\phi \in (0,b)$ for the KdVE). As it is proved in Section 2.1, under these conditions there exists a soliton-like solution $\overline{u}(x,t) = e^{i\omega_0 t}\phi(\omega_0, x)$ for the NLSE (resp., $\overline{u}(x,t) = \phi(\omega_0, x - \omega_0 t)$ for the KdVE) vanishing as $x \to \pm\infty$, for which there exists $\frac{\partial}{\partial\omega}\phi(\omega, \cdot)\big|_{\omega=\omega_0} \in L_2$ and*

$$\frac{d}{d\omega}P(\phi(\omega, \cdot))\Big|_{\omega=\omega_0} = 2 \int_{-\infty}^{\infty} \phi(\omega_0, x)\phi_\omega'(\omega_0, x)dx.$$

*If the condition $\frac{dP(\phi(\omega,\cdot))}{d\omega}\Big|_{\omega=\omega_0} > 0$ is satisfied, then this soliton-like solution $\overline{u}(x,t)$ is stable.*

<u>Remark III.1.7</u> A similar statement is obtained in the paper [92] for soliton-like solutions with positive functions $\phi$ of a multidimensional NLSE. However, as the author of this paper notes, his proof is incomplete when $N > 1$ except for several cases. For example, in the case $N > 1$ and $f(x, |u|^2) = |u|^p$ the proof from [92] is complete. As we know, for any $\omega > 0$ there exists a solitary wave $\overline{u}(x,t) = e^{i\omega t}\phi(\omega, x)$ ($\overline{u}(x,t) = \phi(\omega, x - \omega t)$ for the KdVE), where $\phi(x) > 0$ and, according to results of the paper [35], $\phi$ is radial about a point from $R^N$, if $p \in (0, \frac{4}{N-2})$ ($N > 2$) and $p > 0$ ($N = 1, 2$). The function $\phi$ satisfies equation (II.0.1). Substituting $\phi(x) = \omega^{\frac{1}{p}}v(\omega^{\frac{1}{2}}x)$,

we find in the case of the NLSE:

$$\Delta v - v + |v|^p v = 0, \quad v|_{|x|\to\infty} = 0$$

and by analogy for the KdVE. Hence, by the uniqueness of a positive radial solution of the problem (II.0.1),(II.0.3) mentioned in Additional remarks to Chapter 2, we have $P(\overline{u}) = \omega^{\frac{2}{p}-\frac{N}{2}} \int\limits_{R^N} v^2(y)dy$ where $v(\cdot)$ is a fixed function. Thus, $\frac{dP(\overline{u})}{d\omega} > 0$ if $0 < p < \frac{4}{N}$. According to Theorem III.1.6 (for $N = 1$) and results from the paper [92] (for $N > 1$) the solution $\overline{u}(x,t)$ is stable under this condition. By analogy, in the case of the KdVE with $f(u) = |u|^p$ the condition $\frac{dP(\phi)}{d\omega} > 0$ is satisfied if $p \in (0,4)$. In what follows, we shall show that the instability takes place for the NLSE if $p > \frac{4}{N}$. A similar result on the instability when $p > 4$ for the KdVE is proved in [15].

<u>Proof of Theorem III.1.6.</u>  We first consider the NLSE. We can accept that $f(0) = 0$ and, respectively, $\omega_0 > 0$ making the change of variables $v(x,t) = e^{-if(0)t}u(x,t)$ if necessary, where $u(x,t)$ is a solution of the NLSE (III.1.3). Here we also accept that $||h||_1^2 = \int\limits_{-\infty}^{\infty} [|h'(x)|^2 + \omega_0|h(x)|^2]dx$. One can prove that the greatest lower bound in the expression for $d(\overline{u}, u)$ is achieved at some $\tau = \tau(t) \in R$ and $\gamma = \gamma(t) \in R$ (we remark that generally $\tau$ and $\gamma$ are not unique). We represent a perturbed solution $u$ in the form $e^{-i(\gamma(t)+\omega_0 t)}u(\cdot + \tau(t), t) = \phi + h(x,t)$ where $h(x,t) = v + iw$ ($v,w$ are real). Differentiating the expression for $d(\overline{u}, u)$ with respect to $\tau$ and $\gamma$, we get

$$\int\limits_{-\infty}^{\infty} v[f(\phi^2) + 2\phi^2 f'(\phi^2)]\phi'_x dx = 0, \tag{III.1.5}$$

$$\int\limits_{-\infty}^{\infty} w\phi f(\phi^2)dx = 0. \tag{III.1.6}$$

Further,

$$\Delta E + \frac{\omega_0}{2}\Delta P = E(u) - E(\phi) + \frac{\omega_0}{2}[P(\phi+h) - P(\phi)] \geq \frac{1}{2}\{(L_+ v, v) + (L_- w, w)\} + \alpha(||h||_1^2) \tag{III.1.7}$$

where $\alpha(s) = o(s)$ as $s \to +0$ and $L_+ = -\frac{d^2}{dx^2} + \omega_0 - [f(\phi^2) + 2\phi^2 f'(\phi^2)]$, $L_- = -\frac{d^2}{dx^2} + \omega_0 - f(\phi^2)$.

<u>Lemma III.1.8</u> *There exists $C > 0$ such that $(L_- w, w) \geq C||w||_1^2$ for all $w \in H^1$ satisfying (III.1.6).*

Proof. Let $w = a\phi + w_\perp$ where $(\phi, w_\perp) = 0$. Then, since $\phi$ is the eigenfunction of the operator $L_-$ corresponding to the eigenvalue $\lambda_1 = 0$, we have

$$(L_- w, w) = (L_- w_\perp, w_\perp) \geq \lambda_2 |w_\perp|_2^2, \tag{III.1.8}$$

where $\lambda_2$ is the greatest lower bound of the positive spectrum of $L_-$; $\lambda_2$ is positive because, since $\phi$ is a positive function, the corresponding eigenvalue $\lambda_1 = 0$ is minimal and, hence, is an isolated point of the spectrum of the operator $L_-$ (on this subject, see [28]). Then, by (III.1.6):

$$a \int \phi^2 f(\phi^2) dx + \int w_\perp \phi f(\phi^2) dx = 0.$$

Since $\int\limits_{-\infty}^{\infty} \phi^2 f(\phi^2) dx = \int\limits_{-\infty}^{\infty} (\phi'^2 + w_0 \phi^2) dx > 0$, we get:

$$|a| \le C |w_\perp|_2,$$

hence $|w|_2 \le C_1 |w_\perp|_2$, and therefore (III.1.8) implies $(L_- w, w) \ge C_2 |w|_2^2$ with some $C_2 > 0$ independent of $w \in H^1$. For $k > 0$ independent of $w$ we have ($M = \sup\limits_{x} |f(\phi^2)|$) :

$$(L_- w, w) = \left\{ \frac{1}{k+1}(|w'|_2^2 + w_0 |w|_2^2) - \int\limits_{-\infty}^{\infty} w^2 f(\phi^2) dx \right\} + \frac{k}{k+1}(|w'|_2^2 + w_0 |w|_2^2).$$

Thus,

$$(L_- w, w) \ge \frac{k}{k+1} \|w\|_1^2$$

for $k > 0$ sufficiently small and independent of $w$, because for a sufficiently small $k > 0$ the expression $\frac{1}{k+1}(|w'|_2^2 + w_0 |w|_2^2) - \int\limits_{-\infty}^{\infty} w^2 f(\phi^2) dx$ is not smaller than

$$\frac{C_2}{k+1} |w|_2^2 - \frac{k}{k+1} \int\limits_{-\infty}^{\infty} w^2 f(\phi^2) dx \ge \left( \frac{C_2}{k+1} - \frac{k}{k+1} M \right) |w|_2^2 \ge 0.$$

Lemma III.1.8 is proved. □

In what follows, we use the condition

$$(v, \phi) = 0. \tag{III.1.9}$$

<u>Lemma III.1.9</u> *There exists $C > 0$ such that $(L_+ v, v) \ge C \|v\|_1^2$ for all $v \in H^1$ satisfying (III.1.5) and (III.1.9).*

Proof. Since clearly $\phi'$ is an eigenfunction of $L_+$ with the corresponding eigenvalue $\lambda_2 = 0$ and $\phi'$ has precisely one root, $\lambda_2$ is the second eigenvalue of $L_+$ so that there exists an eigenfunction $g_1(x) > 0$ with a corresponding eigenvalue $\lambda_1 < 0$ and this eigenvalue is minimal. Let $g_1 > 0$, $g_2 = m\phi'$ be eigenfunctions of the operator

$L_+$ normalized in $L_2$ and let $L_\perp$ be the subspace of $L_2$ orthogonal to $g_1$, $g_2$. Let $\phi'_\omega = ag_1 + bg_2 + \phi_\perp$, $v = kg_1 + lg_2 + v_\perp$ where $\phi_\perp$, $v_\perp \in L_\perp$. Then,

$$(L_+v, v) = \lambda_1 k^2 + (L_+v_\perp, v_\perp). \tag{III.1.10}$$

It follows from the spectral theorem (see [28]) that

$$(L_+v_\perp, v_\perp) \geq C_1|v_\perp|_2^2 \tag{III.1.11}$$

where $C_1 > 0$ is independent of $v$. Further, equality (II.0.1) implies that $\phi = -L_+\phi'_\omega$. Therefore, by (III.1.9)

$$0 = -(\phi, v) = \lambda_1 ak + (L_+\phi_\perp, v_\perp).$$

Hence, using the Schwartz inequality, we obtain

$$(L_+v_\perp, v_\perp)^{\frac{1}{2}}(L_+\phi_\perp, \phi_\perp)^{\frac{1}{2}} \geq |(L_+v_\perp, \phi_\perp)| \geq -\lambda_1|a|\,|k|. \tag{III.1.12}$$

Using now conditions of the theorem, we find

$$0 < \frac{d}{d\omega}P(\phi)\Big|_{\omega=\omega_0} = 2(\phi, \phi'_\omega) = -2(L_+\phi_\omega, \phi_\omega) = -2\lambda_1 a^2 - 2(L_+\phi_\perp, \phi_\perp).$$

Hence, there exists $r \in (0, 1)$ independent of $v$ such that

$$0 \leq (L_+\phi_\perp, \phi_\perp) \leq -r\lambda_1 a^2. \tag{III.1.13}$$

Combining (III.1.10)-(III.1.13), we get the following inequality:

$$(L_+v, v) = \lambda_1 k^2 + (L_+v_\perp, v_\perp) \geq \lambda_1 k^2 + \frac{(\lambda_1|a|\,|k|)^2}{-\sqrt{r}\lambda_1 a^2} + (1 - \sqrt{r})(L_+v_\perp, v_\perp) =$$

$$= (1 - \frac{1}{\sqrt{r}})\lambda_1 k^2 + (1 - \sqrt{r})(L_+v_\perp, v_\perp) \geq C_2|kg_1 + v_\perp|_2^2, \quad C_2 > 0. \tag{III.1.14}$$

Now, let us use (III.1.5). Substituting $v = kg_1 + lg_2 + v_\perp$ into this equality, we obtain:

$$|lm| \int_{-\infty}^{\infty} [\phi'_x]^2\{f(\phi^2) + 2\phi^2 f'(\phi^2)\}dx = |lm| \int_{-\infty}^{\infty} [(\phi''_{xx})^2 + \omega_0(\phi'_x)^2]dx =$$

$$= \left| \int_{-\infty}^{\infty} \phi'_x[f(\phi^2) + 2\phi^2 f'(\phi^2)](kg_1 + v_\perp)dx \right| \leq C_3|kg_1 + v_\perp|_2.$$

Thus, $|l| \leq C_4|kg_1 + v_\perp|_2$. This inequality together with (III.1.14) implies:

$$(L_+v, v) \geq C_5|v|_2^2, \quad C_5 > 0.$$

Proceeding further as at the end of the proof of Lemma III.1.8, from the last inequality we get:

$$(L_+v, v) \geq C||v||_1^2, \quad C > 0,$$

and Lemma III.1.9 is proved.□

Lemma III.1.10 $||h(\cdot, t)||_1$ *is a continuous function of* $t$.
Proof. We have for arbitrary $t_1$ and $t_2$:

$$||h(\cdot, t_1)||_1 - ||h(\cdot, t_2)||_1 =$$

$$= ||u(\cdot, t_1) - e^{i(\gamma(t_1) + \omega_0 t_1)}\phi(\cdot - \tau(t_1))||_1 - ||u(\cdot, t_2) - e^{i(\gamma(t_2) + \omega_0 t_2)}\phi(\cdot + \tau(t_2))||_1 \leq$$

$$\leq ||u(\cdot, t_1) - e^{i(\gamma(t_2) + \omega_0 t_2)}\phi(\cdot - \tau(t_2))||_1 - ||u(\cdot, t_2) - e^{i(\gamma(t_2) + \omega_0 t_2)}\phi(\cdot - \tau(t_2))||_1 \leq$$

$$\leq ||u(\cdot, t_1) - u(\cdot, t_2)||_1.$$

By analogy, $||h(\cdot, t_2)||_1 - ||h(\cdot, t_1)||_1 \leq ||u(\cdot, t_2) - u(\cdot, t_1)||_1$, hence

$$|\ ||h(\cdot, t_1)||_1 - ||h(\cdot, t_2)||_1\ | \leq ||u(\cdot, t_1) - u(\cdot, t_2)||_1,$$

and Lemma III.1.10 is proved.□

Now we can prove Theorem III.1.6. Let $h = a\phi + h_\perp$ where $a = a_1 + ia_2$ and $(\phi, h_\perp) = 0$. Then, the functional

$$\Delta P = P(\phi + h) - P(\phi) = 2a_1|\phi|_2^2 + |h|_2^2$$

is independent of $t$; hence

$$|a_1(t) - a_1(0)| = C_1\big|\ |h(\cdot, t)|_2^2 - |h(\cdot, 0)|_2^2\ \big|. \tag{III.1.15}$$

Further, by (III.1.7) and Lemmas III.1.8 and III.1.9

$$\Delta E + \frac{\omega_0}{2}\Delta P \geq C_2\big(||\mathrm{Im}\ h_\perp + a_2\phi||_1^2 + ||\mathrm{Re}\ h_\perp||_1^2\big) - C_3|a_1|\ ||\mathrm{Re}\ h_\perp||_1 - C_4 a_1^2 + \alpha(||h||_1^2),$$

where $\lim_{s \to +0} \frac{\alpha(s)}{s} = 0$. Hence, there exists $m > 0$ such that

$$\Delta E + \frac{\omega_0}{2}\Delta P \geq C_3\big(||\mathrm{Im}\ h_\perp + a_2\phi||_1^2 + ||\mathrm{Re}\ h_\perp||_1^2\big), \quad C_3 > 0, \tag{III.1.16}$$

for all functions $h$ sufficiently small in $H^1$ with corresponding coefficients $a_1$ satisfying the condition $|a_1| < m(||\mathrm{Im}\ h_\perp + a_2\phi||_1 + ||\mathrm{Re}\ h_\perp||_1)$.
Also, clearly

$$\Delta E + \frac{\omega_0}{2}\Delta P \leq C_4||h||_1^2, \quad C_4 > 0, \tag{III.1.17}$$

for all $h$ sufficiently small in $H^1$. For an arbitrary $\delta > 0$ let $O_\delta$ be the open neighborhood of zero in $H^1$ of the kind:

$$O_\delta = \left\{ h \in H^1 : \ |a_1| < \frac{1}{2}\delta, \quad \|\operatorname{Im} h_\perp + a_2\phi\|_1 + \|\operatorname{Re} h_\perp\|_1 < \delta \right\}.$$

Let $\overline{C} > m^{-1}$ be a sufficiently large constant. Let us prove that

$$\sup_{t>0,\ u_0 \in H^1 :\ h(\cdot,0) \in O_\delta} \{|a_1(t)| + \|\operatorname{Im} h_\perp(\cdot,t) + a_2\phi(\cdot,t)\|_1 + \|\operatorname{Re} h_\perp(\cdot,t)\|_1\} \to 0$$

$$\text{(III.1.18)}$$

as $\delta \to +0$. Suppose this is not right. Then, there exist a (sufficiently small) $\epsilon > 0$ and a sequence $u_n(\cdot,t)$ of solutions of the problem (III.1.3),(III.1.4) with $h_n(\cdot,0) \in O_{\delta_n}$, $\delta_n \to +0$ as $n \to \infty$, such that either $a_{1n}(t_n) \geq \epsilon$ or $\|\operatorname{Im} h_{\perp n}(\cdot,t_n) + a_{2n}(t_n)\phi(\cdot,t_n)\|_1^2 + \|\operatorname{Re} h_{\perp n}(\cdot,t_n)\|_1^2 \geq \overline{C}^2\epsilon^2$ for some $t_n > 0$, $n = 1,2,3,\ldots$. First of all, (III.1.15) implies that for all sufficiently large $n$ we have $|a_{1n}(t)| < \epsilon$ until $\|\operatorname{Im} h_{\perp n}(\cdot,t) + a_{2n}(t)\phi(\cdot,t)\|_1^2 + \|\operatorname{Re} h_{\perp n}(\cdot,t)\|_1^2 \leq \overline{C}^2\epsilon^2$ because if $|a_{1n}(t)| = \epsilon$ and $\|\operatorname{Im} h_{\perp n}(\cdot,t) + a_{2n}(t)\phi(\cdot,t)\|_1^2 + \|\operatorname{Re} h_{\perp n}(\cdot,t)\|_1^2 \leq \overline{C}^2\epsilon^2$, then by (III.1.15) $|a_{1n}(t)| \leq \frac{\delta_n}{2} + C_5\epsilon^2$ which is a contradiction because $\epsilon > 0$ is arbitrary small. Hence, according to our assumption, there exist $t_n > 0$, $n = 1,2,3,\ldots$, such that

$$|a_{1n}(t_n)| < \epsilon \quad \text{and} \quad \|\operatorname{Im} h_{\perp n}(\cdot,t_n) + a_{2n}(t_n)\phi(\cdot,t_n)\|_1^2 + \|\operatorname{Re} h_{\perp n}(\cdot,t_n)\|_1^2 = \overline{C}^2\epsilon^2.$$

Then, we have for all sufficiently large numbers $n$:

$$|a_{1n}(t_n)| < m[\|\operatorname{Im} h_{\perp n}(\cdot,t_n) + a_{2n}(t_n)\phi(\cdot,t_n)\|_1 + \|\operatorname{Re} h_{\perp n}(\cdot,t_n)\|_1],$$

therefore, according to (III.1.16), taking a sufficiently large $\overline{C} > m^{-1}$ and a sufficiently small $\epsilon > 0$ independent of $n$, we get

$$\left[ \Delta E + \frac{\omega_0}{2}\Delta P \right]\Big|_{t=t_n} \geq C_6 > 0$$

for all sufficiently large numbers $n$. At the same time, from (III.1.17) we get

$$\left[ \Delta E + \frac{\omega_0}{2}\Delta P \right]\Big|_{t=t_n} = \left[ \Delta E + \frac{\omega_0}{2}\Delta P \right]\Big|_{t=0} \leq C_7\delta_n^2 \to 0 \quad \text{as } n \to \infty.$$

Thus, we arrive at a contradiction, and relations (III.1.18) are proved for all $t$ for which the solution $u(\cdot,t)$ exists. These relations also yield a priori estimates

$$\|u(\cdot,t)\|_1 \leq C_8, \quad t > 0,$$

taking place for all solutions $u(x,t)$ of the problem (III.1.3),(III.1.4) sufficiently close to the soliton-like solution $\overline{u}(\cdot,t)$ at the point $t = 0$ in the sense of the distance $d$ and for those $t > 0$ for which these solutions exist. Hence, according to Theorem I.2.4 and proved statements, any of these solutions $u(\cdot, t)$ is global (it can be continued onto the entire half-line $t > 0$). Thus, for the case of the NLSE the statement of Theorem III.1.6 is proved.

Let us now consider the case of the KdVE. We can assume, making an appropriate change of variables, that $f(0) = 0$ and $\omega_0 > 0$. Representing a perturbed solution $u(x, t)$ in the form $u(x + \tau, t) = \phi(\omega_0, x) + h(x, t)$, where $\overline{u}(x, t) = \phi(\omega_0, x - \omega_0 t)$ is the soliton-like solution under consideration and the parameter $\tau = \tau(t)$ is chosen to minimize

$$|u_x'(\cdot + \tau, t) - \phi_x'(\omega_0, \cdot)|_2^2 + \omega_0 |u(\cdot + \tau, t) - \phi(\omega_0, \cdot)|_2^2,$$

we get the constraint

$$\int_{-\infty}^{\infty} f(\phi(\omega_0, x)) \phi_x'(\omega_0, x) h(x, t) dx = 0. \qquad \text{(III.1.19)}$$

Also, as in the case of the NLSE

$$\Delta E_1 + \frac{\omega_0}{2} \Delta P = E_1(u(\cdot, t)) - E_1(\phi(\omega_0, \cdot)) + \frac{\omega_0}{2} [P(u(\cdot, t)) - P(\phi(\omega_0, \cdot))] \geq$$

$$\geq (Lh, h) + \gamma(\|h\|_1),$$

where $\frac{\gamma(s)}{s^2} \to 0$ as $s \to +0$ and $L = -\frac{d^2}{dx^2} + \omega_0 - f(\phi(\omega_0, \cdot))$. Proceeding further as in the case of the NLSE, we get the estimate

$$(Lh, h) \geq C_9 \|h\|_1^2$$

for all $h \in H^1$ satisfying (III.1.19) and the condition

$$\int_{-\infty}^{\infty} h(x) \phi(\omega_0, x) dx = 0.$$

The end of the proof of the stability in this case repeats those for the NLSE. Theorem III.1.6 is proved.□

**Remark III.1.11** Under the assumptions of Theorem III.1.6 in both cases of the NLSE and KdVE the instability with respect to the above distances takes place if $\frac{dP(\phi)}{d\omega} < 0$. In the case of the KdVE it is proved in the paper [15]; for the NLSE, in the papers [40,84]. Here we consider these questions only for the NLSE and, to simplify our consideration, we focus on the case $f(|u|^2) = |u|^p$. According to results from Chapter 2, the problem (II.0.1),(II.0.3) with $\omega > 0$ has a positive solution if and only if $0 < p < \frac{4}{N-2}$ (to be interpreted as $p > 0$ for $N = 1, 2$); in addition, if $p$ satisfies these restrictions, then the problem has a radial positive solution $\phi$. According to

Remark III.1.7, we have to prove the instability of solutions $e^{i\omega t}\phi(x)$ with respect to the distance $d$ for $p \in \left(\frac{4}{N}, \frac{4}{N-2}\right)$ where $N \geq 3$ ($p > \frac{4}{N}$ for $N = 1, 2$). We present only a formal (incomplete) proof.

Let us use the notation from Section 1.3. First of all, for a given number $\omega = \omega_0 > 0$ and a positive radial solution $\phi$ of the problem (II.0.1),(II.0.3) there exist positive constants $a$ and $b$ such that

$$|\phi(x)| + |\nabla\phi(x)| \leq ae^{-b|x|}$$

for all $x$ as in Section 2.1. Then, $z(t) \equiv 0$ for the solitary wave solution $\bar{u}(x,t) = e^{i\omega_0 t}\phi(\omega_0, x)$ of the NLSE. Further, one can easily verify that for any $\epsilon > 0$ there exists a complex function $h = h_1 + ih_2 \in H^1$ such that the following inequalities take place:

$$\|h\|_1 < \epsilon, \quad |y_{\phi+h}(0) - y_\phi(0)| < \epsilon, \quad |E(\phi+h) - E(\phi)| < \epsilon \quad \text{and} \quad z_{\phi+h}(0) > \epsilon^{-1},$$
$$\text{(III.1.19)}$$

where $y_{\phi+h}$ and $z_{\phi+h}$ are values of the functions $y$ and $z$ corresponding to the solution $e^{it\omega_0}\phi(\omega_0, x) + h(x,t)$ of the Cauchy problem (III.1.3),(III.1.4) with $u_0 = \phi + h$. Thus, we can choose a function $h$ arbitrary small in $H^1$ such that at $t = 0$ the right-hand side of (I.2.19) is not smaller than 1. Further, it is clear that $y_{\phi+h}(t)$ is a decreasing function and $z_{\phi+h}(t)$ is an increasing function of $t$ for all $t > 0$ for which these functions are determined. Therefore, inequality (I.2.19) takes place for all these values of $t > 0$. Hence, the function $z_{\phi+h}(t)$ cannot be continued for all $t > 0$ and there exists $T > 0$ such that $z_{\phi+h}(t) \to +\infty$ as $t \to T - 0$. Hence, $\sup_{t\in[0,T]} |\nabla u|_2 = +\infty$ where $u(\cdot, t) = e^{it\omega_0}\phi(\omega_0, x) + h(x,t)$ and the instability of the solution $e^{i\omega_0 t}\phi(\omega_0, x)$ of the NLSE is proved.$\square$

## 3.2   Stability of kinks for the KdVE

In this section, we shall prove the stability of kinks of the KdVE. We recall that kinks are travelling waves $\phi(\omega, x - \omega t)$ satisfying the conditions $\phi'_x(\omega, x) \neq 0$ and $|\phi(\omega, x)| \leq C$, $x \in R$. We set $\phi(\omega, \pm\infty) = \phi_\pm$. As it is noted in the Introduction, the solutions of this kind are "almost always" stable. It means that they are stable under natural assumptions of the general type on the function $f(u)$. Although we consider only the KdVE, there is a common main idea in the analysis of the stability of kinks for various "soliton" equations; this idea first appeared in the paper by D.B. Henry, J.F. Perez and W.F. Wreszinski [42] where a semilinear wave equation is considered. Here we exploit this idea, too. It is used for estimates from below of the functionals like $(h, Lh)$ (see in what follows).

Consider the KdVE (III.1.1). Conditions providing the existence of kinks are considered in Section 2.1: for a kink $\phi(\omega, x - \omega t)$ satisfying equation (II.1.1) and the conditions $\phi(\pm\infty) = \phi_{\pm}$ to exist, it is sufficient and necessary that the follow-ing conditions are satisfied (here $\overline{f}(\phi) = \int_{\phi_-}^{\phi} f(s)ds$, $f_1(\phi) = \overline{f}(\phi) - \omega\phi + \omega\phi_-$ and $F_1(\phi) = \int_{\phi_-}^{\phi} f_1(s)ds$):

    A: $f_1(\phi_-) = f_1(\phi_+) = 0$;
    B: $F_1(\phi_-) = F_1(\phi_+) = 0$;
    C: $F_1(\phi) < 0$ for all $\phi \in (\phi_-, \phi_+)$.

We also require

$$-\omega + f(\phi_{\pm}) < 0. \tag{III.2.1}$$

Clearly, condition (III.2.1) provides the estimates

$$|\phi(\omega, x) - \phi_{\pm}| + |\phi'_x(\omega, x)| \le C_1 e^{-C_2 |x|}, \quad C_1, C_2 > 0.$$

Without the loss of generality we accept that $\phi_+ > \phi_-$. Under conditions A-C and (III.2.1) we shall show the stability of the kink $\phi(\omega, x - \omega t)$. For this aim we need a suitable result on the existence and uniqueness of a solution $u(x,t)$ of the Cauchy problem (III.1.1),(III.1.2) with conditions on the infinity $u(\pm\infty, t) = \phi_{\pm}$. This result is the following.

Theorem III.2.1 *Let the assumptions A-C and (III.2.1) be valid, $f(\cdot)$ be a twice continuously differentiable function and a function $u_0(\cdot)$ be such that $u_0(\cdot) - \phi(\omega, \cdot) \in H^2$. Then, there exist a half-interval $[0, a)$ and a unique solution $u(x, t)$ of the problem (III.1.1),(III.1.2) such that $u(\cdot, t) - \phi(\omega, \cdot) \in C([0, a); H^2) \cap C^1([0, a); H^{-1})$. For any of these solutions the quantity*

$$I(u(\cdot, t)) = \int_{-\infty}^{\infty} \left\{ \frac{1}{2} u_x^2(x, t) - F_1(u(x, t)) \right\} dx$$

*does not depend on t, i. e. the functional $I(\cdot)$ is a conservation law. In addition, if the above solution exists on a half-interval $[0, a)$, $a > 0$, and there exists $C > 0$ such that $\|u(\cdot, t) - \phi(\omega, \cdot)\|_1 < C$ for all $t \in [0, a)$, then there exists a (unique) continuation of this solution onto a half-interval $[0, a + \delta)$, $\delta > 0$.*

The <u>Proof</u> of this theorem can be made by analogy with the proof of Theorem I.1.3 and Proposition III.1.1.

Remark III.2.2 To get a careful definition of solutions $u(x, t)$ from Theorem III.2.1, it suffices to write the equation for the difference $u - \phi$ and formulate a definition analogous to I.1.1.

Remark III.2.3 Since by construction $|F_1(u)| \leq C(u - \phi_\pm)^2$ as $x \to \pm\infty$, the quantity $I(u(\cdot, t))$ is well-defined. A formal verification of its independence of $t$ is also obvious.

The result on the stability of kinks for the KdVE we consider here is the following.

Theorem III.2.4 *Let $f(\cdot)$ be twice continuously differentiable. Let also conditions A-C and (III.2.1) be satisfied and $\phi(\omega, x - \omega t)$ be the corresponding kink of the KdVE. For any $u(\cdot)$ such that $u(\cdot) - \phi(\omega, \cdot) \in H^1$ we set*

$$\rho_q^2(u, \phi) = \inf_{\tau \in R} \{|u_x'(x) - \phi_x'(\omega, x - \tau)|_2^2 + q|u(x) - \phi(\omega, x - \tau)|_2^2\},$$

*where $q = \max_{s \in [\phi_-, \phi_+]} \{|f(s)| + |\omega| + 1\}$. Then, the kink $\phi$ is stable with respect to the distance $\rho_q$, i. e. for any $\epsilon > 0$ there exists $\delta > 0$ such that if $u_0(\cdot) - \phi(\omega, \cdot) \in H^2$ and $\rho_q(u_0(\cdot), \phi(\omega, \cdot)) < \delta$, then the corresponding solution $u(\cdot, t)$ of the problem (III.1.1), (III.1.2) from Theorem III.2.1 can be continued onto the entire half-line $t > 0$ and for any $t > 0$ the inequality $\rho_q(u(\cdot, t), \phi(\omega, \cdot)) < \epsilon$ takes place.*

Remark III.2.5 One can easily see that the stability from Theorem III.2.4 is again the stability of the form.

Proof of Theorem III.2.4. We first prove the following estimate:

$$I(u) - I(\phi) \geq C\rho_q^2(u, \phi) - \alpha(\rho_q^2(u, \phi)), \tag{III.2.2}$$

where $\alpha(s) = o(s)$ as $s \to +0$ and $C > 0$ is independent of $u$. Let $u(x, t) = \phi(\omega, x - \tau - \omega t) + h(x, t)$ where $\tau = \tau(t) \in R$ is chosen for $\rho_q(u, \phi)$ to be minimal (one can easily prove the existence of such $\tau$). Then,

$$\Delta I = I(u) - I(\phi) =$$

$$= \frac{1}{2} \int_{-\infty}^{\infty} \{h_x^2 + [\omega - f(\phi)]h^2\} \, dx - \frac{1}{2} \int_{-\infty}^{\infty} \{f(\phi + \theta h) - f(\phi)\} h^2 dx = \frac{1}{2} I_1(h) + \alpha(\rho_q^2(u, \phi)),$$

$$\tag{III.2.3}$$

where $\theta = \theta(x, t) \in (0, 1)$, $\alpha(s) = o(s)$ as $s \to +0$ and $I_1(h) = \int_{-\infty}^{\infty} hLh dx$ with $L = -\frac{d^2}{dx^2} + \omega - f(\phi)$.

Now we use the spectral theory of symmetric ordinary differential operators (for details, see [28]). The continuous spectrum of the operator $L$ coincides with the set $[\sigma, +\infty)$ where $\sigma = \omega - \max\{f(\phi_+), f(\phi_-)\} > 0$. Since $\phi'(x) > 0$ and $L\phi' = 0$, the function $\phi'$ is the first (positive) eigenfunction of the operator $L$ with the corresponding minimal eigenvalue $\lambda_1 = 0$. Hence, the second eigenvalue $\lambda_2$ is positive if it exists. We take $b = \lambda_2$ if $\lambda_2$ exists and $b = \sigma$ in the opposite case.

Let $h = \mu\phi' + g$ where $(\phi', g) = 0$. We obviously have

$$I_1(h) \geq b|g|_2^2. \tag{III.2.4}$$

Further, differentiating $\rho_q^2(u, \phi)$ with respect to $\tau$ at the point of minimum, we find

$$\int\limits_{-\infty}^{\infty} [q - \omega + f(\phi)]\phi' h \, dx = 0.$$

Substituting $h = \mu\phi' + g$ into this equality, we easily get

$$\mu \leq C_1|g|_2$$

which together with (III.2.4) gives

$$I_1(h) \geq C_2|h|_2^2, \quad C_2 > 0.$$

At last, we get the estimate

$$(Lh, h) \geq C_3\|h\|_1^2, \quad C_3 > 0, \tag{III.2.5}$$

as in the proof of Lemma III.1.8. Now inequality (III.2.2) easily follows from (III.2.3) and (III.2.5).

The last part of the proof of Theorem III.2.4 is clear. As in Lemma III.1.10, the continuity of $\|u(\cdot, t) - \phi(\omega, \cdot)\|_1$ yields the continuity of $\|h(\cdot, t)\|_1$ as a function of $t$. Take an arbitrary $\epsilon > 0$. Then, in view of the above arguments there exist $\delta_0 > 0$ and $0 < \overline{C} < \overline{\overline{C}}$ such that

$$\overline{C}\|h\|_1^2 \leq \Delta I \leq \overline{\overline{C}}\|h\|_1^2 \tag{III.2.6}$$

for all $u \in H^1 : \rho_q(u, \phi) < \delta_0$. Let $\delta_1 \in (0, \delta_0)$ be such that $\sqrt{\delta_1 \overline{C}^{-1}} < \epsilon$. Take an arbitrary $\delta > 0$ for which $\overline{\overline{C}}\delta^2 < \delta_1$. Let $u(\cdot, 0)$ be such that $\|h(\cdot, t)\|_1 < \delta$. Then, due to (III.2.6) we have $\Delta I < \delta_1$, therefore, in view of the continuity of $\|h(\cdot, t)\|_1$ in $t$, we get

$$\|h(\cdot, t)\|_1 < \sqrt{\delta_1 \overline{C}^{-1}} < \epsilon \tag{III.2.7}$$

for all $t > 0$ for which the solution $u(\cdot, t)$ exists. Thus, by the latter statement of Theorem III.2.1 and since due to (III.2.7)

$$\|u(\cdot, t)\|_1 < C'$$

in any finite half-interval $[0, a)$ of the existence of the solution $u(\cdot, t)$, the solution $u(\cdot, t)$ is global in time and Theorem III.2.4 is proved.□

## 3.3 Stability of solutions of the NLSE nonvanishing as $|x| \to \infty$

In this section, we shall consider two results on the stability of solutions of the one-dimensional NLSE (III.1.3) (with $N = 1$) non-vanishing as $x \to \pm\infty$. These solutions are space-homogeneous ones $\Phi(t) = a_0 e^{if(a_0^2)t}$, where $a_0 > 0$, and kinks. As we already noted in the Introduction, we shall prove a stability of a new interesting type. We begin with studying solutions $\Phi(t) = a_0 e^{if(a_0^2)t}$, where $a_0 > 0$ is a parameter.

Let $c \in R$, $u(\cdot) \in X^1$ and $u(x) \neq 0$ for all $x \in R$ (this is valid, for example, if $u(\cdot)$ is sufficiently close in $X^1$ to a nonzero constant such as $\Phi(t)$ with a fixed $t \in R$). Then, one can easily verify that there exist a unique real-valued function $a(\cdot) \in X^1$ and a function $\omega(\cdot)$, absolutely continuous in any finite interval and unique up to adding $2\pi m$, $m = 0, \pm 1, \pm 2, \ldots$ to it, such that

$$u(x) = (a_0 + a(x))e^{i[c+\omega(x)]} \tag{III.3.1}$$

and that $a_0 + a(x) = |u(x)| > 0$ for all $x \in R$; in addition, due to the relation

$$u'(x) = [a'(x) + i(a_0 + a(x))\omega'(x)]e^{i[c+\omega(x)]}, \tag{III.3.2}$$

we have $(a_0+a(\cdot))\omega'(\cdot) \in L_2$. In particular, if $0 < c_1 \leq a_0+a(\cdot) \leq c_2 < +\infty$ for all $x \in R$, that for example occurs if $\|a(\cdot)\|_1$ is sufficiently small, then $\omega'(\cdot) \in L_2$. Conversely, if for a complex-valued function $u(\cdot)$ there exist real-valued functions $a(\cdot) \in X^1$ and $\omega(\cdot)$, absolutely continuous in any finite interval, such that $(a_0 + a(\cdot))\omega'(\cdot) \in L_2$, $a_0 + a(x) > 0$ for all $x \in R$ and that (III.3.1) with some $c \in R$ takes place, then $u(\cdot) \in X^1$ and $u(x) \neq 0$ for all $x \in R$. The result on the stability of solutions $\Phi(t)$ we consider here is the following.

Theorem III.3.1 *Let $N = 1$, $f(|u|^2)u$ be a twice continuously differentiable function of the complex argument $u$, $f(\cdot)$ be a real-valued function and $f'(a_0^2) < 0$ for some $a_0 > 0$. Let $\delta_0 > 0$ be such that $|a(\cdot)|_C < \frac{a_0}{2}$ if $a(\cdot) \in H^1$ and $\|a(\cdot)\|_1 < \delta_0$. Then, the solution $\Phi(t) = a_0 e^{if(a_0^2)t}$ of the NLSE (III.1.3) is stable in the following sense: for an arbitrary $\epsilon \in (0, \delta_0)$ there exists $\delta \in (0, \delta_0)$ such that, if $u_0(\cdot) \in X^1$, $u_0(x) \neq 0$ and $u_0(x) = (a_0 + \bar{a}(x))e^{i\omega(x)}$ where $\bar{a}(x) = |u_0(x)| - a_0 \in H^1$, $\|\bar{a}(\cdot)\|_1 < \delta$ and $|\omega'(\cdot)|_2 < \delta$, then the corresponding $X^1$-solution $u(x,t)$ of the Cauchy problem (III.1.3),(III.1.4) given by Theorem I.2.10 can be continued onto the entire half-line $t > 0$, $u(x,t) \neq 0$ for all $x \in R$, $t > 0$ and for any fixed $t > 0$ the functions $a(x,t)$ and $\omega(x,t)$ in the representation*

$$u(x,t) = (a_0 + a(x,t))e^{i[f(a_0^2)+\omega(x,t)]}$$

*satisfy the following: $a(\cdot, t) \in H^1$, $\|a(\cdot, t)\|_1 < \epsilon$ and $|\omega'_x(\cdot, t)|_2 < \epsilon$.*

<u>Proof.</u> Let first $k \geq 1$ be integer and $u_0(\cdot) \in X^k$. If we suppose a sufficient smoothness of the function $f(|u|^2)u$, then, according to Theorem I.2.10, there exists a unique (local) $X^k$-solution of the problem (III.1.3),(III.1.4). Clearly, for any integer $l = 1, 2, ..., k$ this problem has a unique local $X^l$-solution which, of course, coincides with the $X^k$-solution in all intervals of the existence of both solutions. Let positive $T_1, T_2, ..., T_k$ be such that, for any $l = 1, 2, ..., k$, $[0, T_l)$ be the maximal half-interval of the existence of the $X^l$-solution. Clearly, $T_1 \geq T_2 \geq ... \geq T_k > 0$.

<u>Lemma III.3.2</u> *Under the above assumptions* $T_1 = T_2 = ... = T_k = T > 0$.
<u>Proof.</u> Due to the inequality

$$|\frac{d^l}{dx^l}f(|u|^2)u|_2 \leq C'_l(|||u|||_{l-1}) + C''_l(|||u|||_1)|\frac{d^l}{dx^l}u|_2,$$

we have

$$|||u|||_l \leq C_1(\max_{s\in[0,t]}|||u(\cdot,s)|||_{l-1}) + C_2(\max_{s\in[0,t]}|||u|||_1)\int_0^t |||u(\cdot,s)|||_l ds$$

which step by step implies the statement of Lemma III.3.2.□

Let $I(u) = \int\limits_{-\infty}^{\infty} (|u|^2 - a_0^2)^2 dx$.

<u>Lemma III.3.3</u> *Let the assumptions of Theorem III.3.1 be valid and let* $I(u_0) < \infty$ *for some* $u_0 \in X^1$ *and the corresponding* $X^1$*-solution* $u(x,t)$ *of the Cauchy problem (III.1.3),(III.1.4) exist for* $t \in [0, t_0)$. *Then,* $I(u(\cdot,t)) < \infty$ *for all* $t \in [0, t_0)$ *and this function is continuous in* $t$.
<u>Proof.</u> At first let $u_0 \in X^3$ and $f(|u|^2)u$ be a four times continuously differentiable function. For the corresponding $X^3$-solution $u(x,t)$ of the Cauchy problem (III.1.3),(III.1.4) given by Theorem I.2.10 we formally have:

$$\frac{dI(u)}{dt} = \lim_{a,b\to+\infty} \{2i[(|u|^2 - a_0^2)(u_x\bar{u} - u\bar{u}_x)]|_{x=-a}^{x=b} + 2i\int_{-a}^{b} (u^2\bar{u}_x^2 - u_x^2\bar{u}^2))dx\}.$$

Then, according to the standard result, since $I(u(\cdot,0)) < \infty$ and the limit in the right-hand side of the latter equality is uniform with respect to $t$ from an arbitrary closed subset of $[0, t_0)$, $I(u(\cdot,t)) < \infty$ for all $t \in [0, t_0)$ and the function $I(u(\cdot,t))$ is continuously differentiable in $t$. Now the statement of Lemma III.3.3 can be obtained for $u_0 \in X^1$ by the passage to the limit over a sequence of smoothed functions $u_0$ and $f(|u|^2)u$ converging to non-smoothed ones.□

Lemma III.3.4 *Let the assumptions of Theorem III.3.1 be valid, $u_0 \in X^1$ and the corresponding $X^1$-solution of the problem (III.1.3),(III.1.4) exists for $t \in [0, t_0)$. Let*

$$M(u) = \int\limits_{-\infty}^{\infty} \left\{ \frac{1}{2}|u_x|^2 - U(|u|^2) + \frac{f(a_0^2)}{2}|u|^2 + D \right\} dx,$$

*where $D = -\frac{f(a_0^2)}{2}a_0^2 + U(a_0^2)$ and $U(s) = \frac{1}{2}\int\limits_0^s f(p)dp$. If $I(u_0) < \infty$, then the quantity $M(u(\cdot, t))$ is determined for all $t \in [0, t_0)$ and independent of $t$.*

Proof. One can easily verify that, since $I(u_0) < \infty$ and because

$$\left| -U(|u|^2) + \frac{f(a_0^2)}{2}|u|^2 + D \right| \leq C(u)(|u| - a_0)^2,$$

the quantity $M(u_0)$ is determined. Let first $u_0 \in X^3$, $f(|u|^2)u$ be a four times continuously differentiable function, and let $u(x, t)$ be the corresponding $X^3$-solution of the problem (III.1.3),(III.1.4). Then, the formal differentiation shows that $\frac{dM(u(\cdot, t))}{dt} = 0$, hence indeed $M(u(\cdot, t))$ is determined and independent of $t$. For a twice continuously differentiable function $f(|u|^2)u$ and $u_0 \in X^1$ one can easily obtain the statement of Lemma III.3.4 by a passage to the limit over sequences of smoothed functions $u_0$ and $f(|u|^2)u$ as in the proof of Lemma III.3.3. □

Let $u_0 \in X^1$ be such that $I(u_0) < \infty$ and $u_0(x) \neq 0$, $x \in R$. Let also $[0, t_0)$ be the maximal half-interval such that the corresponding $X^1$-solution of the problem (III.1.3),(III.1.4) exists for $t \in [0, t_0)$ and is nonzero for these values of $t$ and $x \in R$. Then, in view of Lemmas III.3.3 and III.3.4 for $t \in [0, t_0)$ we can rewrite the functional $M(u)$ in the following form:

$$M(u(\cdot, t)) = M(u(\cdot, t)) - M(\Phi) = M_1 + M_2 + M_3,$$

where $M_1(u(\cdot, t)) = \frac{1}{2}\int\limits_{-\infty}^{\infty} \{a_x^2(x, t) - 2a^2(x, t)a_0^2 f'(a_0^2)\}dx$, $M_2(u(\cdot, t)) =$

$$= \frac{1}{2} \int\limits_{-\infty}^{\infty} \omega_x^2(x, t)(a_0 + a(x, t))^2 dx, \quad M_3(u(\cdot, t)) = \alpha(\|a(\cdot, t)\|_1^2)$$

and $\lim\limits_{s \to +0} \frac{\alpha(s)}{s} = 0$ (here $u(x, t) = (a_0 + a(x, t))e^{i[f(a_0^2)t + \omega(x,t)]}$ where $a_0 + a(x, t) > 0$ and for any fixed $t$ as functions of $x$ the function $a(\cdot, t)$ belongs to $H^1$ and the function $\omega(x, t)$ is absolutely continuous in an arbitrary finite interval).

By Lemma III.3.3 $\|a(\cdot, t)\|_1$ is a continuous function of $t \in [0, t_0)$. Also, in view of the condition of the theorem $f'(a_0^2) < 0$, there exists $C > 0$ such that

$$M_1(u) \geq C\|a\|_1^2, \quad a \in H^1. \tag{III.3.3}$$

Take an arbitrary $\epsilon \in (0, \delta_0)$ and let $H(s) = Cs^2 + \alpha(s^2)$ where $C > 0$ is the constant from the estimate (III.3.3). Then, there exists $\delta_1 \in (0, \epsilon)$ such that $H(s) \geq \frac{C}{2}s^2$ for any $s \in (0, \delta_1]$. By the above arguments, there exists $\delta_2 \in (0, \delta_1)$ such that

$$H(\|a(\cdot, t)\|_1) \leq M(u) < \frac{C}{4}\delta_1^2$$

if $\|a(\cdot, 0)\|_1 < \delta_2$ and $|\omega_x'(\cdot, 0)|_2 < \delta_2$. Then, using the continuity of $\|a(\cdot, t)\|_1$, we get for such initial data $\|a(\cdot, t)\|_1 < \delta_1 < \epsilon$ for all $t \in [0, t_0)$.

At last, in view of the equality

$$\frac{1}{2}|\omega_x(a_0 + a)|_2^2 = M(u) - M_1(u) - M_3(u)$$

we get $|\omega_x'(\cdot, t)|_2 < \epsilon$ if $\|a(\cdot, 0)\|_1 < \delta_2$ and $|\omega_x'(\cdot, 0)|_2 < \delta_2$ where $\delta_2 > 0$ is sufficiently small. Therefore, for these initial data by (III.3.2) there exists $C_1 > 0$ such that $\||u(\cdot, t)\||_1 < C_1$ for all $t \in [0, t_0)$. Thus, $t_0 = +\infty$, and Theorem III.3.1 is proved.$\Box$

**Corollary III.3.5** *Let the assumptions of Theorem III.3.1 be valid. Then, the solution $\Phi(t)$ is also stable in the following sense: for any $\epsilon \in (0, \delta_0)$ and (an arbitrary large) $d > 0$ there exists $\delta \in (0, \delta_0)$ such that if $u_0 \in X^1$, $|u_0(\cdot)| - a_0 \in H^1$, $\| |u_0(\cdot)| - a_0\|_1 < \delta_0$ (consequently $u_0(x) = (a_0 + \bar{a}(x))e^{i\omega(x)}$ where $\omega(\cdot)$ is a real-valued function absolutely continuous in any finite interval and $\omega'(\cdot) \in L_2$) and for the corresponding $X^1$-solution $u(x, t)$ of the problem (III.1.3), (III.1.4) given by Theorem I.2.10 $\|a(\cdot, 0)\|_1 < \delta$ and $|\omega_x'(\cdot, 0)|_2 < \delta$, then for any $t > 0$ and $x_0 \in R$ there exists $\gamma = \gamma(x_0, t)$ such that*

$$\|u(\cdot, t)e^{i\gamma} - \Phi(t)\|_{H^1(x_0-d, x_0+d)} < \epsilon.$$

Proof immediately follows from the proved Theorem III.3.1.$\Box$

**Remark III.3.6** The assumptions of Theorem III.3.1 are valid, for example, for $f(s) = -s$. According to this theorem, the corresponding solutions $\Phi(t) = a_0e^{-ia_0^2 t}$, $a_0 > 0$, are stable.

Now we consider the stability of kinks for the NLSE. Let $f(|u|^2)u$ be a continuously differentiable function of the complex argument $u$, $f(\cdot)$ be a real-valued function and let $\bar{\omega}$, $\phi_-$ and $\phi_+$ be such that the following conditions are satisfied:
   (a) $\phi_-, \phi_+ > 0$;
   (b) $-\bar{\omega} + f(\phi_\pm^2) + 2\phi_\pm^2 f'(\phi_\pm^2) < 0$;
   (c) $-\bar{\omega} + f(\phi_\pm^2) = 0$;
   (d) $-\frac{\bar{\omega}}{2}\phi_-^2 + U(\phi_-^2) = -\frac{\bar{\omega}}{2}\phi_+^2 + U(\phi_+^2)$ where $U(s) = \frac{1}{2}\int\limits_0^s f(r)dr$;

(e) $-\frac{\overline{\omega}}{2}s^2 + U(s^2) < -\frac{\overline{\omega}}{2}\phi_-^2 + U(\phi_-^2)$ for all $s \in (\phi_-, \phi_+)$.

As it is shown in Section 2.1, conditions (c)-(e) provide the existence of a kink $\overline{u}(x,t) = e^{i\overline{\omega}t}\phi(x)$ of the NLSE (III.1.3) with $\phi(\pm\infty) = \phi_\pm$. As it is well-known, condition (b) implies the existence of $c > 0$ such that

$$\lim_{x \to -\infty} [\phi(x) - \phi_-]e^{-cx} = \lim_{x \to +\infty} [\phi(x) - \phi_+]e^{cx} = \lim_{x \to -\infty} \phi'(x)e^{-cx} = \lim_{x \to +\infty} \phi'(x)e^{cx} = 0.$$

The assumption (a) is technical; we shall need it in what follows. We also suppose for the definiteness that $\phi_- < \phi_+$.

Let a function $g(x)$ be such that $g(\cdot) - \phi(\cdot) \in H^1$. We denote by $\tau_0$ a real number minimizing the function $\|g(\cdot - \tau) - \phi(\cdot)\|_1$ (since $\phi(\cdot - \tau) - \phi(\cdot) \in H^1$, we have $g(\cdot - \tau) - \phi(\cdot) = g(\cdot - \tau) - \phi(\cdot - \tau) + \phi(\cdot - \tau) - \phi(\cdot) \in H^1$). Of course, as earlier, generally speaking, $\tau_0$ is non-unique.

Let a $X^1$-solution $u(x,t)$ of the NLSE (III.1.3) be such that $|u(\cdot,t)| - \phi(\cdot) \in H^1$ for some $t \geq 0$. We set $v(x,t) = u(x - \tau_0, t)$, where $\tau_0$ is taken for $g(x) = |u(x,t)|$, and $a(x,t) = |v(x,t)| - \phi(x)$. As earlier, if $\|a(\cdot,t)\|_1$ is sufficiently small, then $0 < c_1 \leq \phi(x) + a(x,t) \leq c_2 < \infty$ for this value $t$ and respectively there exists a real-valued function $\omega(x,t)$, absolutely continuous in an arbitrary finite interval and unique up to adding the term $2\pi m$, $m = \pm 1, \pm 2, ...$, to it, such that $v(x,t) = (\phi(x) + a(x,t))e^{i[\overline{\omega}t + \omega(x,t)]}$. Since $v(\cdot,t) \in X^1$ and $v'_x(x,t) = [\phi'(x) + a'_x(x,t) + i(\phi(x) + a(x,t))\omega'_x(x,t)]e^{i[\overline{\omega}t + \omega(x,t)]}$, we have $\omega'_x(\cdot,t) \in L_2$ if $\|a(\cdot,t)\|_1$ is sufficiently small. By analogy, if $u(x,t) \neq 0$ for some $t \geq 0$ and all $x \in R$, where $u(x,t)$ is a $X^1$-solution of the problem (III.1.3),(III.1.4), then we can introduce $\omega(x,t)$.

<u>Theorem III.3.7</u> *Let $N = 1$, $f(|u|^2)u$ be a twice continuously differentiable function of the complex argument $u$ and let the assumptions (a)-(e) be valid. Then, the corresponding kink $\overline{u}(x,t) = e^{i\overline{\omega}t}\phi(x)$ of the NLSE (III.1.3) is stable in the following sense: for any sufficiently small $\epsilon > 0$ there exists a sufficiently small $\delta > 0$ such that if $u_0 \in X^1$, $|u_0(\cdot)| - \phi(\cdot) \in H^1$, $\|a(\cdot,0)\|_1 < \delta$ and $|\omega'_x(\cdot,0)|_2 < \delta$, then the corresponding $X^1$-solution $u(x,t)$ of the problem (III.1.3),(III.1.4) is global (it can be continued onto the entire half-line $t > 0$) and for any fixed $t > 0$ one has $|u(\cdot,t)| - \phi(\cdot) \in H^1$, $\|a(\cdot,t)\|_1 < \epsilon$ and $|\omega'_x(\cdot,t)|_2 < \epsilon$.*

<u>Proof.</u> Consider the functional

$$M(u) = \int_{-\infty}^{\infty} \left\{ \frac{1}{2}|u_x(x)|^2 - U(|u(x)|^2) + \frac{\overline{\omega}}{2}|u(x)|^2 + D_1 \right\} dx,$$

where $D_1 = -\frac{\overline{\omega}}{2}\phi_-^2 + U(\phi_-^2)$. One can easily verify that for the quantity $M(u(\cdot))$ to be determined for a function $u(\cdot) \in X^1$ it is sufficient and necessary that $|u(\cdot)| - \phi(\cdot) \in L_2$.

<u>Lemma III.3.8</u> *Let $u(x,t)$ be a $X^1$-solution of the problem (III.1.3),(III.1.4) in an interval of time $[0, t_0)$ and let $|u(\cdot,0)| - \phi(\cdot) \in L_2$. Then, $|u(\cdot,t)| - \phi(\cdot) \in L_2$ for*

all $t \in [0, t_0)$ so that the quantity $M(u(\cdot, t))$ is determined for all $t \in [0, t_0)$ and, in addition, independent of $t$.

Proof. Let first $f(|u|^2)u$ be a five times continuously differentiable function, $u_0 \in X^4$ and $u(x, t)$ be the corresponding $X^4$-solution of the problem (III.1.3),(III.1.4). Then, the direct calculation shows that $\frac{dM(u(\cdot, t))}{dt} = 0$, hence, the quantity $M(u(\cdot, t))$ is determined and independent of $t \in [0, t_0)$. For a $X^1$-solution we get the statement of the lemma by a passage to the limit over a sequence of smoothed functions $f(|u|^2)u$ and $u_0$ converging to the non-smoothed ones.□

For $u_0 \in X^1$ nonequal to zero and such that $|u_0(\cdot)| - \phi(\cdot) \in H^1$ we have

$$\Delta M = M(u_0) - M(\phi) = M(v) - M(\phi) = M_1 + M_2 + M_3, \qquad \text{(III.3.4)}$$

where

$$M_1 = \frac{1}{2} \int_{-\infty}^{\infty} \{a_x^2 + (\overline{\omega} - f(\phi^2) - 2\phi^2 f'(\phi^2))a^2\}dx, \quad M_2 = \frac{1}{2} \int_{-\infty}^{\infty} w_x'^2(\phi + a)^2 dx,$$

$$M_3 = \frac{1}{2} \int_{-\infty}^{\infty} a^2\{f(\phi^2) + 2\phi^2 f'(\phi^2) - f((\phi + \theta a)^2) - 2(\phi + \theta a)^2 f'((\phi + \theta a)^2)dx \quad \text{(III.3.5)}$$

where $\theta = \theta(x, t) \in (0, 1)$.

Lemma III.3.9 There exists $C_1 > 0$ such that $M_1(g) \geq C_1 |g|_2^2$ for all real-valued $g \in H^1$ satisfying the condition

$$(g, \phi') = 0. \qquad \text{(III.3.6)}$$

Proof. Consider the operator $L\varphi = -\varphi'' - q(x)\varphi$ where $q(x) = -\overline{\omega} + f(\phi^2) + 2\phi^2 f'(\phi^2)$. Let $q_\pm = -\overline{\omega} + f(\phi_\pm^2) + 2\phi_\pm^2 f'(\phi_\pm^2)$. By condition (b) we have $q_\pm < 0$. The continuous spectrum of the operator $L$ fills the half-line $[b, +\infty)$ where $b = \min\{q_-; q_+\} > 0$ (see [28]). Further, it follows from equation (III.1.3) that $\phi'(x)$ is an eigenfunction of the operator $L$ with the corresponding eigenvalue $\lambda_0 = 0$; in addition, since $\phi'(x)$ is of constant sign, $\lambda_0$ is the smallest and isolated eigenvalue of the operator $L$. Hence, it follows from the spectral theorem that $M_1(g) = (g, Lg) \geq C_1 |g|_2^2$ for all $g \in H^1$ satisfying (III.3.6), and Lemma III.3.9 is proved.□

The number $\tau_0$ was defined as a point of minimum of the function $R(\tau) = \| |u(\cdot - \tau, t)| - \phi(\cdot)\|_1$. Therefore,

$$0 = (R^2(\tau))' \Big|_{\tau=\tau_0} = 2 \int_{-\infty}^{\infty} a(x, t)\phi'(x)[q - \overline{\omega} + f(\phi^2) + 2\phi^2 f'(\phi^2)]dx.$$

We take $\bar{q} = \sup\limits_{x \in R} |\bar{\omega} - f(\phi^2(x)) - 2\phi^2(x)f'(\phi^2(x))| + 1$ and

$$K = \frac{|\phi'|_2 \; |\phi'(\bar{q} - \bar{\omega} + f(\phi^2) + 2\phi^2 f'(\phi^2))|_2}{\int\limits_{-\infty}^{\infty} \phi'^2(\bar{q} - \bar{\omega} + f(\phi^2) + 2\phi^2 f'(\phi^2))dx}.$$

**Lemma III.3.10** $M_1(g) \geq C_2|g|_2^2$, where $C_2 = C_1(1 + K)^{-2}$, for all real-valued $g \in H^1$ satisfying the condition

$$\int\limits_{-\infty}^{\infty} g(x)\phi'(x)(\bar{q} - \bar{\omega} + f(\phi^2(x)) + 2\phi^2(x)f'(\phi^2(x)))dx = 0. \qquad \text{(III.3.7)}$$

**Proof.** Represent an arbitrary function $g \in H^1$ satisfying (III.3.7) in the form $g = \alpha\phi' + \varphi$ where $(\phi', \varphi) = 0$. Then $M_1(g) = M_1(\varphi) \geq C_1|\varphi|_2^2$. We get from condition (III.3.7):

$$\alpha \int\limits_{-\infty}^{\infty} \phi'^2[\bar{q} - \bar{\omega} + f(\phi^2) + 2\phi^2 f'(\phi^2)]dx + \int\limits_{-\infty}^{\infty} \varphi\phi'[\bar{q} - \bar{\omega} + f(\phi^2) + 2\phi^2 f'(\phi^2)]dx = 0,$$

hence,

$$|\alpha| \; |\phi'|_2 = |\phi'|_2 \left| \frac{\int\limits_{-\infty}^{\infty} \varphi\phi'[\bar{q} - \bar{\omega} + f(\phi^2) + 2\phi^2 f'(\phi^2)]dx}{\int\limits_{-\infty}^{\infty} \phi'^2[q - \omega + f(\phi^2) + 2\phi^2 f'(\psi^2)]du} \right| \leq$$

$$\leq |\varphi|_2 \; |\phi'|_2 \frac{|\phi'[\bar{q} - \bar{\omega} + f(\phi^2) + 2\phi^2 f'(\phi^2)]|_2}{\int\limits_{-\infty}^{\infty} \phi'^2[\bar{q} - \bar{\omega} + f(\phi^2) + 2\phi^2 f'(\phi^2)]dx} = K|\varphi|_2.$$

In addition $|g|_2 \leq |\alpha| \; |\phi'|_2 + |\varphi|_2 \leq (1 + K)|\varphi|_2$. Thus $M_1(g) \geq C_1|\varphi|_2^2 \geq C_2|g|_2^2$, and Lemma III.3.10 is proved. $\square$

**Lemma III.3.11** There exists $C_3 > 0$ such that $M_1(g) \geq C_3\|g\|_1^2$ for all real-valued $g \in H^1$ satisfying (III.3.7).

**Proof.** The inequality $M_1(g) \geq C_2|g|_2^2$ from Lemma III.3.10 implies the following inequality with an arbitrary $k > 0$:

$$M_1(g) \geq \frac{k}{2(1+k)}|g'_x|_2^2 + \frac{C_2}{(1+k)}|g|_2^2 + \frac{k}{2(1+k)}\int\limits_{-\infty}^{\infty} q(x)g^2(x)dx.$$

Hence,

$$M_1(g) \geq \frac{k}{2(1+k)}|g'_x|_2^2 + \frac{2C_2 - k\bar{q}}{2(1+k)}|g|_2^2.$$

Thus, taking a sufficiently small $k > 0$, we get $M_1(g) \geq C_3\|g\|_1^2$ with some $C_3 > 0$.□

**Lemma III.3.12** *There exists a nondecreasing function $L(s)$ of the argument $s \geq 0$ such that $\lim_{s \to +0} L(s) = 0$ and $|M_3(g)| \leq \|g\|_1^2 L(\|g\|_1)$.*
Proof is obvious.□

**Lemma III.3.13** *For any $\epsilon > 0$ there exists a sufficiently small $\delta > 0$ such that $|M(v(\cdot, 0)) - M(\phi)| < \epsilon$ for any nonzero $u_0(\cdot) \in X^1$ satisfying the conditions $|u_0(\cdot)| - \phi(\cdot) \in H^1$, $\|a(\cdot, 0)\|_1 < \delta$ and $|\omega'(\cdot, 0)|_2 < \delta$.*
Proof follows from Lemma III.3.12 and relations (III.3.4) and (III.3.5).□

**Lemma III.3.14** *Let $u_0 \in X^1$ be such that $a(\cdot, 0) = |v(\cdot, 0)| - \phi(\cdot) \in H^1$ and the corresponding $X^1$-solution $u(x, t)$ of the problem (III.1.3), (III.1.4) exist in a half-interval of time $[0, t_0)$. Then, $a(\cdot, t) = |v(\cdot, t)| - \phi(\cdot) \in H^1$ for all $t \in [0, t_0)$ and $\|a(\cdot, t)\|_1$ is a continuous function of $t$.*
Proof. It follows from the formal inequalities

$$\|a(\cdot, t_1)\|_1 - \|a(\cdot, t_2)\|_2 = \||\phi(\cdot + \tau(t_1)) - |u(\cdot, t_1)|\,\|_1 - \||\phi(\cdot + \tau(t_2)) - |u(\cdot, t_2)|\,\|_1 \leq$$

$$\leq \||\phi(\cdot + \tau(t_2)) - |u(\cdot, t_1)|\,\|_1 - \||\phi(\cdot + \tau(t_2)) - |u(\cdot, t_2)|\,\|_1 \leq \|\,|u(\cdot, t_1)| - |u(\cdot, t_2)|\,\|_1$$

and Lemma III.3.8 that it suffices to prove that $|u(\cdot, t_1)| - |u(\cdot, t_2)| \in H^1$ for any $t_1, t_2 \in [0, t_0)$ and that $\lim_{t_2 \to t_1} \|\,|u(\cdot, t_2)| - |u(\cdot, t_1)|\,\|_1 = 0$. The first claim follows from the fact that $|u(\cdot, t)| - \phi(\cdot) \in H^1$. To prove the second, consider the expression $J(u) = \int_{-\infty}^{\infty} [|u(x)|^2 - \phi^2(x)]^2 dx$. Clearly, if $J(u(\cdot, t))$ is determined for $t \in [0, t_0)$ and continuous in $t$, then the second claim holds.

If $f(|u|^2)u$ is a sufficiently smooth function and $u_0 \in X^3$, then for the corresponding $X^3$-solution $u(x, t)$ we have as in the proof of Lemma III.3.3 that $J(u(\cdot, t))$ is a continuously differentiable function of $t \in [0, t_0)$ and

$$\frac{dJ(u(\cdot, t))}{dt} = 2i \int_{-\infty}^{\infty} [(u^2\bar{u}_x^2 - u_x^2\bar{u}^2) + 2\phi\phi_x(u\bar{u}_x - u_x\bar{u})]dx.$$

For an arbitrary $X^1$-solution $u(\cdot, t)$ of the problem (III.1.3), (III.1.4) satisfying the condition $a(\cdot, 0) \in H^1$ we get the same property and relation by passing to the limit over a sequence of smoothed solutions.□

Let us prove Theorem III.3.7. Let $[0, t_0)$ be the maximal interval of the existence of a $X^1$-solution of the problem (III.1.3), (III.1.4) satisfying $|u(\cdot, 0)| - \phi(\cdot) \in H^1$,

$||a(\cdot,0)||_1$ be sufficiently small and let $H(s) = C_3 s^2 - s^2 L(s)$. Then, due to (III.3.4), (III.3.5) and Lemmas III.3.11 and III.3.12

$$\Delta M(t) \geq H(||a(\cdot,t)||_1). \tag{III.3.8}$$

By Lemma III.3.12 there exists $\delta_1 > 0$ such that $H(s) \geq \frac{C_3}{2} s^2$ for all $s \in (0, \delta_1)$ and that $|g(\cdot)|_C < \frac{\phi_-}{2}$ for all $g(\cdot) \in H^1$ satisfying $||g(\cdot)||_1 < \delta_1$. Take an arbitrary $\epsilon \in (0, \delta_1)$ and apply Lemma III.3.13, according to which there is $\delta_2 \in (0, \epsilon)$ such that $\Delta M < \frac{C_3}{4} \epsilon^2$ if $||a(\cdot,0)||_1 < \delta_2$ and $|\omega_x'(\cdot,0)|_2 < \delta_2$. Then, by Lemmas III.3.11 and III.3.14 and inequality (III.3.8) $||a(\cdot,t)||_1 < \epsilon$ for all $t \in [0, t_0)$ if $||a(\cdot,0)||_1 < \delta_2$ and $|\omega_x'(\cdot,0)|_2 < \delta_2$. Observe also that

$$0 \leq |\omega_x'(\cdot,t)|_2 \leq C_4 (\Delta M - M_1 - M_3) \to 0$$

uniformly in $t \in [0, t_0)$ as $||a(\cdot,0)||_1 + |\omega_x'(\cdot,0)|_2 \to 0$. Thus there exists $\overline{C} > 0$ such that $|||u(\cdot,t)|||_1 \leq \overline{C}$ for all $t \in [0, t_0)$ for each solution $u(x,t)$ of the problem (III.1.3),(III.1.4) sufficiently close to the kink at the point $t = 0$ in the above sense, i. e. these solutions are global. Theorem III.3.7 is proved.□

**Remark III.3.15** Theorem III.3.7 was first proved in the paper [106] where also the case $0 = \phi_- < \phi_+$ is considered. If $\phi_- < 0 < \phi_+$, then our method of proving the theorem does not work since it is impossible to represent an arbitrary perturbed solution $u(x,t)$ in the form $u(x,t) = (\phi(x) + a(x,t))e^{i[\overline{\omega}t + \omega(x,t)]}$ because a perturbed solution can have no root but the right-hand side of this representation has a root $x = x(t)$ if $a(\cdot,t) \in H^1$.

**Remark III.3.16** Now we want to comment Theorems III.3.1 and III.3.7. First of all, we observe that the smallness of $|\omega_x'(\cdot,t)|_2$ for a fixed $t$ does not mean that $\omega(x,t)$ as a function of $x$ is close to a constant, it can be even unbounded as one can easily verify. At the same time, as in the Corollary III.3.5, for any $\epsilon > 0$ and $d > 0$ there exists $\delta > 0$ such that, if $|\omega_x'(\cdot,t)|_2 < \delta$, then $|\omega(x_1,t) - \omega(x_2,t)| < \epsilon$ for any $x_1$ and $x_2$ for which $|x_1 - x_2| < 2d$ so that the function $\omega(x,t)$ is close to a constant in any interval $(x_0 - d, x_0 + d)$ of the length $2d$ tending to $+\infty$ as $|\omega_x'(\cdot,t)|_2 \to 0$, but the constants, the function $\omega(x,t)$ is close to, can be different for intervals $(x_1 - d, x_1 + d)$ and $(x_2 - d, x_2 + d)$, if the difference $|x_1 - x_2|$ is sufficiently large. Thus, roughly speaking, in each case of Theorem III.3.1 or III.3.7 the structure of a perturbed solution sufficiently close to the unperturbed one at $t = 0$ is close to the structure of the unperturbed solution in any above bounded interval but phases of the perturbed solution can be shifted from one another in two such intervals if these intervals are sufficiently far from each other.

## 3.4 Additional remarks

Partially results presented in this chapter are also contained in the monographs [43,88] and in the review [95]. In [14], some small corrections to the paper by T.B. Benjamin [7] are made. Also, in [8] the stability of travelling waves periodic in the spatial variable $x$ of the KdVE is considered. In the case of the NLSE the distance $d(\cdot, \cdot)$ was first introduced in [21,23] and, independently, in [100].

In Section 3.1 we have obtained a sufficient condition for the stability of soliton-like solutions vanishing as $|x| \to \infty$ called the "Q-criterion". Probably this result was first rigorously proved in the paper [92] (it appeared earlier like many other results on the stability of solitary waves without a rigorous proof in the physical literature). We have studied only the one-dimensional case $N = 1$. In fact, a similar statement for the multidimensional NLSE is used in physical literature (see, for example, [61]). However, as it is already noted, in the multidimensional case such a result is rigorously proved only in several particular situations; the difficulties are partially related to our insufficient knowledge about the stationary problem (II.0.1),(II.0.3) (in particular, to investigate the stability, we need the uniqueness of a positive solution of the latter problem).

Another approach to the investigation of the stability of soliton-like solutions vanishing as $|x| \to \infty$ is based on the concentration-compactness method by P.L. Lions. Although we have considered an application of this method in a simplest case, currently it finds applications to essentially more difficult problems, for example, multidimensional (and sometimes nonlocal). However, it should be noted that attempts to apply this method to the problem of the stability of solitary waves sometimes meet difficulties similar to those which occur in the case when we deal with the "Q-criterion" and related, for example, again to the problem of the uniqueness of stationary solutions.

Another problem we almost do not touch upon in this chapter is the instability. We remind that, roughly speaking, the inequality $\frac{dP(\overline{u})}{d\omega} > 0$ provides the stability of a soliton-like solution $\overline{u}$. In the case when we have the opposite inequality $\frac{dP(\overline{u})}{d\omega} < 0$, the instability is proved in [15] for the KdVE and in [40,84] for the NLSE though proofs from these papers again meet difficulties in the multidimensional case. We refer readers to this literature.

Several papers (see, for example, [22,81]), in which the concentration-compactness method is not used, are devoted to the investigation of the stability of solitary waves for a multidimensional NLSE with coefficients depending on the spatial variable.

As it is known to the author, there are almost no results on the stability of solitary waves of the NLSE non-vanishing as $|x| \to \infty$. In Section 3.3 we have presented two results in this direction.

Finally, we mention the papers [12,40,84,93] where the problem of the stability of solitary waves for the equations under consideration is reduced to abstract problems of the stability in abstract spaces with further applications of the obtained results to the original equations.

# Chapter 4

# Invariant measures

Invariant measures are the basic tool in the theory of dynamical systems. In the case when a dynamical system is finite-dimensional (more precisely, when the phase space of a dynamical system is a compact metric space) the classical result by N.N. Bogoliubov and N.M. Krylov states the existence of a nonnegative normalized invariant measure (see [13,52] and, also, [72]). By analogy, as it is well known, a finite-dimensional Hamiltonian system possesses a Liouville invariant measure. Another situation occurs when the phase space of a dynamical system is infinite-dimensional. Of course, there is a natural interest whether invariant measures exist in this case, too. Numerous recent papers are devoted to proving the existence or constructing such measures in the infinite-dimensional case when a dynamical system is generated by a nonlinear partial differential equation. In this chapter, we illustrate these investigations for the NLSE and KdVE with our two results. Some other literature on this subject is indicated in Additional remarks.

Now we mention some applications of invariant measures we consider. First, these objects are used in [55,65,67] for constructing statistical mechanics of nonlinear wave equations (in [55], without a proof of the invariance). Second, a bounded invariant measure for a dynamical system generated by a nonlinear partial differential equation allows to give an explanation (partial) of the Fermi-Pasta-Ulam phenomenon well-known in the theory of nonlinear waves. For a description of this phenomenon in detail we refer readers to the paper [73]. Briefly, these authors considered a chain of mass points on the circle where each mass point interacts with the neighboring mass points by a nonlinear law. Roughly speaking, they observed (by computer simulations) that the configuration of this system becomes arbitrary close to the initial at some moments of time $t_n \to \infty$. Later, a similar recurrence of trajectories was found (mainly again by computer simulations) for many "soliton" equations with suitable boundary conditions and it was called the Fermi-Pasta-Ulam phenomenon. An analogous property of trajectories of a dynamical system is called the stability according to Poisson. If one has a bounded (nonnegative) invariant measure for a corresponding dynamical system, then the Poincaré recurrence theorem (see below) explains this phenomenon partially.

Now we present a definition of a dynamical system. In the literature, there are different definitions of this concept and we choose the following.

Definition IV.0.1 *Let $M$ be a complete separable metric space and let a function $g : R \times M \longmapsto M$ for any fixed $t$ be a homeomorphism of the space $M$ into itself satisfying the properties:*
*1) $g(0, x) = x$ for any $x \in M$;*
*2) $g(t, g(\tau, x)) = g(t + \tau, x)$ for any $t, \tau \in R$ and $x \in M$.*
*Then, we call the function $g$ a dynamical system with the phase space $M$. If $\mu$ is a Borel measure defined on the phase space $M$ and $\mu(\Omega) = \mu(g(\Omega, t))$ for an arbitrary Borel set $\Omega \subset M$ and for all $t \in R$, then it is called an invariant measure for the dynamical system $g$.*

This definition is sufficient for our goals. In particular, with it the Poincaré recurrence theorem is applicable. For definitions of Borel sets and measures, see the next section. Also, Proposition IV.1.1 provides the correctness of our definition of an invariant measure.

Definition IV.0.2 *Let $h(t, \cdot)$ be a dynamical system with a phase space $M$. Then, we say that a trajectory $h(t, x)$, where $x \in M$ is fixed and $t$ runs over the whole real line $R$, is positively (or negatively) stable according to Poisson if there exists a sequence $\{t_n\}_{n=1,2,3,\ldots}$ tending to $+\infty$ as $n \to \infty$ such that $h(t_n, x) \to x$ (resp. $h(-t_n, x) \to x$) in $M$ as $n \to \infty$. We say that a trajectory is stable according to Poisson if it is positively and negatively stable according to Poisson.*

In this chapter, speaking about applications of constructed invariant measures, we concentrate our attention on recurrence properties of trajectories of corresponding dynamical systems, generated by the KdVE or the NLSE, following from the Poincaré recurrence theorem. We call a nonnegative Borel measure $\mu$ in a metric space $M$ *bounded* if $\mu(M) < \infty$.

Poincaré recurrence theorem *Let $h$ be a dynamical system, the phase space of which is a metric space $M$, and let $\mu$ be a nonnegative bounded invariant measure for this dynamical system. Then, almost all points of $M$ (in the sense of the measure $\mu$) are stable according to Poisson.*

This result is well known. We refer readers to the numerous literature for its proof (see, for example, the monograph [72]).

## 4.1 On Gaussian measures in Hilbert spaces

In this section, we mainly consider only those properties of Gaussian measures which are used in the next sections. We refer readers to the monographs [27,85] and the review paper [32] for an advanced theory of measures in infinite-dimensional spaces. All the information from the general measure theory to be used further is contained for example in the book [41]. Here we briefly recall some basic facts of this theory.

Let $X$ be an infinite set and $\mathcal{A}$ be a set of some subsets of $X$. We call $\mathcal{A}$ an *algebra* if the following two conditions are satisfied:

1. $\emptyset, X \in \mathcal{A}$;
2. $A \cup B$ and $A \setminus B$ belong to $\mathcal{A}$ if $A, B \in \mathcal{A}$.

We call an algebra $\mathcal{A}$ a *sigma-algebra* if $A = \bigcap_{n=1}^{\infty} A_n \in \mathcal{A}$ for any $A_n \in \mathcal{A}$ $(n = 1, 2, 3, ...)$ such that $A_1 \supset A_2 \supset ... \supset A_n \supset ....$ It is known that if $\mathcal{A}$ is a sigma-algebra, then for any $A_n \in \mathcal{A}$ $(n = 1, 2, 3, ...)$ one has $\bigcup_{n=1}^{\infty} A_n \in \mathcal{A}$ and $\bigcap_{n=1}^{\infty} A_n \in \mathcal{A}$. It is known, too, that for any algebra $\mathcal{A}$ there exists a unique sigma-algebra $\mathcal{M}$ containing $\mathcal{A}$ and contained in any sigma-algebra containing $\mathcal{A}$; this sigma-algebra is called the *minimal sigma-algebra containing $\mathcal{A}$*.

A nonnegative function $\nu$ defined on an algebra $\mathcal{A}$ and satisfying $\nu(A \cup B) = \nu(A) + \nu(B)$ for any $A, B \in \mathcal{A}$ satisfying $A \cap B = \emptyset$ is called an *additive measure* defined on $\mathcal{A}$. If $\mu$ is a nonnegative additive measure on $\mathcal{A}$ such that $\lim_{n\to\infty} \mu(A_n) = 0$ for any $A_n \in \mathcal{A}$ $(n = 1, 2, 3, ...)$ for which $A_1 \supset A_2 \supset ... \supset A_n \supset ...$ and $\bigcap_{n=1}^{\infty} A_n = \emptyset$, then the measure $\mu$ is called *countably additive* on the algebra $\mathcal{A}$. If a measure $\mu$ is countably additive on an algebra $\mathcal{A}$, then there exists its unique extension onto the minimal sigma-algebra $\mathcal{M}$ containing $\mathcal{A}$ such that this measure is countably additive on this sigma-algebra in the sense that $\mu\left(\bigcap_{n=1}^{\infty} A_n\right) = \lim_{n\to\infty} \mu(A_n)$ for an arbitrary $A_n \in \mathcal{M}$ $(n = 1, 2, 3, ...)$ satisfying $A_1 \supset A_2 \supset ... \supset A_n \supset ...$ and that $\mu\left(\bigcup_{n=1}^{\infty} A_n\right) = \sum_{n=1}^{\infty} \mu(A_n)$ for any $A_n \in \mathcal{M}$ $(n = 1, 2, 3, ...)$ such that $A_i \cap A_j = \emptyset$ if $i \neq j$.

Let $M$ be a separable complete metric space. Then, the minimal sigma-algebra containing all open and closed subsets of $M$ is called the *Borel sigma-algebra* in $M$; it should be remarked that in general the set of all open and closed subsets of $M$ is not an algebra so that a careful definition of the Borel sigma-algebra needs an additional consideration. A countably additive nonnegative measure defined on the Borel sigma-algebra is called a Borel measure in the space $M$. If $\mu$ is a Borel measure in the space $M$, then for any set $A \in \mathcal{M}$

$$\mu(A) = \sup_{K \subset A} \mu(K) = \inf_{O \supset A} \mu(O),$$

where the supremum is taken over all closed sets $K$ contained in $A$ and the infimum is taken over all open sets $O$ containing $A$.

Now we present the following statement providing, in particular, the correctness of the definition of an invariant measure of a dynamical system.

Proposition IV.1.1 *Let $M$ and $N$ be complete separable metric spaces and $\varphi :$ $M \longmapsto N$ be a homeomorphism of the space $M$ onto $N$. Then, $\varphi$ transforms Borel subsets of the space $M$ into Borel subsets of the space $N$.*

Sketch of the <u>Proof.</u> Let $\mathcal{M}$ and $\mathcal{N}$ denote the Borel sigma-algebras in $M$ and in $N$, respectively. Let $A \in \mathcal{M}$. Suppose that $B = \varphi(A) \notin \mathcal{N}$. Then, all Borel subsets of the space $M$ transforming by $\varphi$ into Borel subsets of the space $N$ obviously form a sigma-algebra $\mathcal{M}_1$, containing all open and closed sets, and $A$ is not contained in $\mathcal{M}_1$. But this is a contradiction in view of the definition of the Borel sigma-algebra. $\square$

Now we briefly consider Gaussian measures in a finite-dimensional space $R^n$. Let $n \geq 1$ be integer, $a \in R^n$ and $B$ be a symmetric positively defined $n \times n$ matrix with real components (we remind that the positive definiteness of a matrix $C$ means that $(Cx, x) > 0$ for any $x \neq 0$ from $R^n$). The Borel nonnegative measure $w$ in $R^n$ with the density

$$\rho(x) = \frac{1}{\sqrt{(2\pi)^n \det (B)}} \exp\left\{-\frac{1}{2}(B^{-1}(x-a), (x-a))\right\}$$

is called a *(nondegenerate) Gaussian measure* in $R^n$; obviously $w(R^n) = 1$. We call this measure *centered* if $a = 0$.

In what follows, we need some properties of the measure $w$. Let $a = 0$. Then, one can easily verify that, for $x = (x_1, ..., x_n) \in R^n$,

$$\int_{R^n} x_i x_j w(dx) = (B)_{i,j} = b_{i,j}$$

so that for a $n \times n$ matrix $A$

$$\int_{R^n} (Ax, x) w(dx) = \mathrm{Tr}\ (AB),$$

where $\text{Tr}\,(C) = \sum_{i=1}^{n} c_{i,i}$ is the trace of a $n \times n$ matrix $C = (c_{i,j})_{i,j=1,2,\dots,n}$. By analogy

$$\int_{R^n} (Ax, x)^2 w(dx) = [\text{Tr}\,(AB)]^2 + 2\text{Tr}\,(AB)^2$$

for a symmetric $n \times n$ matrix $A$. The following result is in fact taken from [27].

<u>Lemma IV.1.2</u> *Let $a = 0$ and let $A$ be a positively defined symmetric $n \times n$ matrix such that $AB = BA$. Then,*

$$w\left(\{x \in R^n : (Ax, x) \geq 1\}\right) \leq \text{Tr}\,(AB)$$

*and*

$$w\left(\left\{x \in R^n : |(Ax, x) - \text{Tr}\,(AB)| < c\sqrt{\text{Tr}\,(AB)}\right\}\right) \geq 1 - 2c^{-2} \max_{n} \mu_n$$

*where $c > 0$ is arbitrary and $\mu_1, \dots, \mu_n$ are eigenvalues of the symmetric positively defined matrix $AB$.*

<u>Proof.</u> We have

$$w\left(\{x \in R^n : (Ax, x) \geq 1\}\right) \leq \int_{R^n} (Ax, x) w(dx) = \text{Tr}\,(AB).$$

By analogy

$$w\left(\left\{x \in R^n : |(Ax, x) - \text{Tr}\,(AB)| \geq c\sqrt{\text{Tr}\,(AB)}\right\}\right) \leq$$

$$\leq \int_{R^n} \frac{[(Ax, x) - \text{Tr}\,(AB)]^2}{c^2 \text{Tr}\,(AB)} w(dx) = 2c^{-2} \frac{\text{Tr}\,[(AB)^2]}{\text{Tr}\,(AB)} \leq 2c^{-2} \max_{n} \mu_n. \quad \square$$

Now, finishing the discussion of the finite-dimensional case, we present a result on invariant measures for systems of autonomous ordinary differential equations not directly related to Gaussian measures but used in what follows. Consider the following system of ordinary differential equations:

$$\dot{x} = \varphi(x),$$

where $x(t) = (x_1(t), \dots, x_n(t)) : R \longmapsto R^n$ is an unknown vector-function and $\varphi(x) = (\varphi_1(x), \dots, \varphi_n(x)) : R^n \longmapsto R^n$ is a continuously differentiable map. Let $h(t, x)$ be the corresponding function ("dynamical system") from $R \times R^n$ into $R^n$ transforming any $t \in R$ and $x_0 \in R^n$ into the solution $x(t)$, taken at the moment of time $t$, of the above system supplied with the initial data $x(0) = x_0$.

Theorem IV.1.3 *Let $M(x_1, ..., x_n)$ be a continuously differentiable function from $R^n$ into $R$. For the Borel measure in $R^n$*

$$\nu(\Omega) = \int_\Omega M(x)dx$$

*to be invariant for the function $h(t, x)$ in the sense that $\nu(h(t, \Omega)) = \nu(\Omega)$ for any bounded open domain $\Omega$ and for any $t$ from an interval $(-T, T)$ so small that $h(t, \Omega)$ is determined and bounded for all $t \in (-T, T)$, it is sufficient and necessary that*

$$\sum_{i=1}^{n} \frac{\partial[M(x)\varphi_i(x)]}{\partial x_i} = 0$$

*for all $x = (x_1, ..., x_n) \in R^n$.*

Proof. We here follow the monograph [72]. Take an arbitrary bounded open set $\Omega \subset R^n$ and let $(-T, T)$ be an interval from the formulation of the theorem. Let

$$I(t) = \int_{h(t,\Omega)} M(x)dx, \quad t \in (-T, T),$$

and $t, t + \tau \in (-T, T)$. Then,

$$I(t + \tau) = \int_{h(t+\tau,\Omega)} M(x_1', ..., x_n')dx_1'...dx_n'.$$

Clearly, the map $h(\tau, h(t, \Omega))$ from $h(t, \Omega)$ into $h(t + \tau, \Omega)$ is a diffeomorphism, i. e. $x' = (x_1', ..., x_n') = \psi(\tau, x) = (\psi_1(\tau, x), ..., \psi_n(\tau, x))$ where for a fixed $\tau$ the map $\psi : h(t, \Omega) \longmapsto h(t + \tau, \Omega)$ is one-to-one and smooth with the inverse. Hence,

$$I(t + \tau) = \int_{h(t,\Omega)} M(\psi(\tau, x)) \left| \det\left(\frac{\partial(x_1', ..., x_n')}{\partial(x_1, ..., x_n)}\right)\right| dx. \qquad (IV.1.1)$$

Further, we have

$$M(\psi(\tau, x)) = M(x_1 + \tau\left(\frac{\partial\psi_1}{\partial\tau}\right)_{\tau=0} + o(\tau), ..., x_n + \tau\left(\frac{\partial\psi_n}{\partial\tau}\right)_{\tau=0} + o(\tau)),$$

where obviously

$$\left(\frac{\partial\psi}{\partial\tau}\right)_{\tau=0} = \varphi(x).$$

Therefore,

$$M(x') = M(x) + \tau\sum_{i=1}^{n}\varphi_i(x)\frac{\partial M(x)}{\partial x_i} + o(\tau). \qquad (IV.1.2)$$

Also, one can easily verify that

$$\left(\frac{\partial x_i'}{\partial x_j}\right)_{\tau=0} = \delta_{ij}, \quad \text{where } \delta_{ij} = 1 \text{ if } i = j \text{ and } 0 \text{ otherwise,}$$

and

$$\frac{d}{d\tau}\frac{\partial x_i'}{\partial x_j} = \sum_{k=1}^{n} \frac{\partial \varphi_i(x)}{\partial x_k'} \frac{\partial x_k'}{\partial x_j} \quad (i, j = 1, 2, ..., n).$$

Hence,

$$\frac{\partial x_i'}{\partial x_j} = \delta_{ij} + \tau \frac{\partial \varphi_i(x)}{\partial x_j} + o(\tau), \quad i, j = 1, 2, ..., n.$$

This yields

$$\det\left(\frac{\partial(x_1', ..., x_n')}{\partial(x_1, ..., x_n)}\right) = 1 + \tau \sum_{i=1}^{n} \frac{\partial \varphi_i(x)}{\partial x_i} + o(\tau),$$

therefore, due to (IV.1.1) and (IV.1.2),

$$I(t + \tau) = \int_{h(t+\tau,\Omega)} \left\{ M(x) + \tau \left[\sum_{i=1}^{n}\left(\varphi_i \frac{\partial M}{\partial x_i} + M \frac{\partial \varphi_i}{\partial x_i}\right)\right] + o(\tau) \right\} dx$$

and thus

$$I'(t) = \int_{h(t,\Omega)} \left[\sum_{i=1}^{n} \frac{\partial(M(x)\varphi_i(x))}{\partial x_i}\right] dx.$$

The latter relation immediately implies the statement of Theorem IV.1.3.□

Now, we turn to study infinite-dimensional Gaussian measures. In several places of this presentation, we follow the book [27]. First, we need the following definition.

Definition IV.1.4 *Let $B$ be a linear bounded self-adjoint operator in a real separable Hilbert space $H$ and let this operator possess a sequence $\{e_n\}_{n=1,2,3,...}$ of eigenelements forming an orthonormal basis in $H$. Let $Be_n = \lambda_n e_n$, $n = 1, 2, 3, ...$, where $\lambda_n > 0$ for all $n$. Then, $B$ is called an operator of trace class iff $\sum_{n=1}^{\infty} \lambda_n < \infty$.*

Definition IV.1.5 *Let $H$ be a real separable Hilbert space and $B : H \to H$ be a self-adjoint operator with eigenelements $\{e_n\}_{n=1,2,3,...}$ forming an orthonormal basis in $H$. We call a set $M \subset H$ cylindrical if there exist integer $n \geq 1$ and a Borel set $F \subset R^n$ such that*

$$M = \{x \in H : [(x, e_1)_H, ..., (x, e_n)_H] \in F\}. \tag{IV.1.3}$$

For a fixed operator $B$ from Definition IV.1.5 we denote by $\mathcal{A}$ the set of all cylindrical subsets of $H$. One can easily verify that $\mathcal{A}$ is an algebra.

Definition IV.1.6 *Let $H$ be again a real separable Hilbert space, $\{e_n\}_{n=1,2,3,...}$ be an orthonormal basis in $H$ and $B$ be a linear self-adjoint operator in $H$ (generally unbounded) defined by the rule $Be_n = \lambda_n e_n$, where $\lambda_n > 0$ are some numbers (so that $x = \sum\limits_{n=1}^{\infty} x_n e_n \in H$ belongs to the domain of the definition of $B$ if and only if $\sum\limits_{n=1}^{\infty} \lambda_n x_n^2 < \infty$). Then, we call the additive (but in general not countably additive!) measure $w$ defined on the algebra $\mathcal{A}$ by the rule: for $M \in \mathcal{A}$ of the kind (IV.1.3)*

$$w(M) = (2\pi)^{-\frac{n}{2}} \prod_{i=1}^{n} \lambda_i^{-\frac{1}{2}} \int_F e^{-\frac{1}{2}\sum\limits_{i=1}^{n} \lambda_i^{-1} x_i^2} dx_1...dx_n$$

*the centered Gaussian measure in $H$ with the correlation operator $B$.*

Proposition IV.1.7 *Let $B$ be an operator of trace class acting in a real separable Hilbert space $H$. Then, the minimal sigma-algebra $\mathcal{M}$ in the space $H$ containing the algebra of all cylindrical sets is the Borel sigma-algebra.*

Proof. It suffices to prove that an arbitrary closed ball $\overline{B_r(a)}$ with $a \in H$ and $r > 0$ belongs to $\mathcal{M}$. This follows from the relation $\overline{B_r(a)} = \bigcap\limits_{n=1}^{\infty} M_n$ where cylindrical sets $M_n$ are defined as follows:

$$M_n = \{x \in H : \ (x - a, e_1)_H^2 + ... + (x - a, e_n)_H^2 \leq r^2\}, \quad n = 1, 2, 3, ... \square$$

The principal result of the theory of Gaussian measures in Hilbert spaces is the following.

Theorem IV.1.8 *The centered Gaussian measure $w$ from Definition IV.1.6 is countably additive on the algebra $\mathcal{A}$ if and only if $B$ is an operator of trace class.*

Proof. First, let $\sum\limits_{n=1}^{\infty} \lambda_n = +\infty$. Suppose that the measure $w$ is countably additive. Consider first the case $\lambda = \sup\limits_{n} \lambda_n < \infty$. Let $P_n$ be the orthogonal projector in $H$ onto the subspace $H_n = \text{span } \{e_1, ..., e_n\}$ and let $R_n = \text{Tr } (P_n B P_n) = \sum\limits_{i=1}^{n} \lambda_i$. Consider cylindrical sets

$$M_n = \left\{x \in H : \ \big|\|P_n x\|_H^2 - R_n\big| < \alpha\sqrt{R_n}\right\}, \quad \alpha > 0.$$

By the second statement of Lemma IV.1.2 $w(M_n) \geq 1 - 2\lambda\alpha^{-2}$. Taking here $\alpha = 2\sqrt{\lambda}$, we get $w(M_n) \geq \frac{1}{2}$. Therefore, since $R_n \to +\infty$ as $n \to \infty$, in $H$ there exist balls $B_{R_n - \alpha\sqrt{R_n}}(0)$ of the arbitrary large radiuses satisfying $w(B_{R_n - \alpha\sqrt{R_n}}(0)) \leq \frac{1}{2}$ which contradicts the countable additivity of the measure $w$.

Let now $\sup_n \lambda_n = +\infty$ and let $0 < \mu_1 \leq \mu_2 \leq ... \leq \mu_n \leq ...$ be a subsequence of the sequence $\{\lambda_n\}_{n=1,2,3,...}$ of eigenvalues of the operator $B$ satisfying $\lim_{n\to\infty} \mu_n = +\infty$ with corresponding eigenelements $\{g_n\}_{n=1,2,3,...}$. For each integer $n \geq 1$ consider the cylindrical set

$$M_n = \{x \in H : |(x, g_i)_H| \leq n, \ i = 1, 2, ..., a_n\},$$

where $a_n > 0$ is integer. We have

$$w(M_n) = (2\pi)^{-\frac{a_n}{2}} \int_{-n\mu_1^{-\frac{1}{2}}}^{n\mu_1^{-\frac{1}{2}}} dx_1... \int_{-n\mu_{a_n}^{-\frac{1}{2}}}^{n\mu_{a_n}^{-\frac{1}{2}}} e^{-\frac{1}{2}\sum_{i=1}^{a_n} x_i^2} dx_{a_n} \leq \left\{ (2\pi)^{-\frac{1}{2}} \int_{-n\mu_1^{-\frac{1}{2}}}^{n\mu_1^{-\frac{1}{2}}} e^{-\frac{x^2}{2}} dz \right\}^{a_n}.$$

We take integer $a_n > 0$ so large that the right-hand side here is smaller than $2^{-n-1}$ for each $n$ and that $a_n \to +\infty$ as $n \to \infty$. Then, on the one hand $\bigcup_{n=1}^{\infty} M_n = H$ and $w(H) = 1$ (because $H$ is a cylindrical set of the kind (IV.1.3) with $F = R^n$), but on the other hand $w(\bigcup_{n=1}^{\infty} M_n) \leq \sum_{n=1}^{\infty} w(M_n) \leq \frac{1}{2}$, i. e. we get a contradiction, and the first part of the theorem is proved.

Now, let $\sum_{n=1}^{\infty} \lambda_n < +\infty$. Let us prove that the measure $w$ is countably additive on the algebra $\mathcal{A}$ of cylindrical sets. For this aim, let us first prove that for any $\epsilon > 0$ there exists a compact set $K_\epsilon \subset H$ such that $w(M) < \epsilon$ for any cylindrical set $M$ for which $M \cap K_\epsilon = \emptyset$.

Let $b_n > 0$ be such that $b_n \to +\infty$ as $n \to \infty$ and $\lambda = \sum_{n=1}^{\infty} b_n \lambda_n < +\infty$. Take arbitrary $\epsilon > 0$ and $R > 0$. Then, by the first inequality from Lemma IV.1.2 for cylindrical sets $M$ of the kind

$$M = \{x \in H : [(x, e_1)_H, ..., (x, e_n)_H] \in F\},$$

where $\sum_{i=1}^{n} b_i x_i^2 > R^2$ for any $x = (x_1, ..., x_n) \in F$ (i. e. $M \cap E = \emptyset$ where $E = \{x \in H : \sum_{n=1}^{\infty} b_n(x, e_n)_H^2 \leq R^2\}$), we have $w(M) \leq \lambda R^{-2}$ and the set $E$ is obviously relatively compact in $H$. Taking for $K_\epsilon$ the closure $\overline{E}$ in $H$ of the set $E$ with $R > \sqrt{\frac{\lambda}{\epsilon}}$, we get the above compact set.

Now, let $A_1 \supset A_2 \supset ... \supset A_n \supset ...$ be a sequence of cylindrical sets in $H$ and let $\bigcap_{n=1}^{\infty} A_n = \emptyset$. Then, by the known property of Borel measures for any $\epsilon > 0$ there exist closed cylindrical sets $C_n \subset A_n$ $(n = 1, 2, 3, ...)$ of the kind (IV.1.3) such that $w(A_n \setminus C_n) < \epsilon 2^{-n-2}$. Let also $D_n = \bigcap_{k=1}^{n} C_k$. Then, $D_n$ are closed cylindrical sets and

$$w(A_n \setminus D_n) \leq w(\bigcup_{k=1}^{n}(A_k \setminus C_k)) < \frac{\epsilon}{2}.$$

Let also $K_{\frac{\epsilon}{2}}$ be the above-constructed compact set corresponding to $\frac{\epsilon}{2}$ instead of $\epsilon$ and let $E_n = D_n \cap K_{\frac{\epsilon}{2}}$. Clearly, $E_n$ are compact sets in $H$, $E_n \subset A_n$ and $w(A_n \setminus E_n) < \epsilon$ for any $n$. In addition, since $\bigcap\limits_{n=1}^{\infty} A_n = \emptyset$, we have $\bigcap\limits_{n=1}^{\infty} E_n = \emptyset$. But hence, there exists a number $n_0 > 0$ such that $E_n = \emptyset$ for all $n \geq n_0$. Thus, $w(A_n) \leq w(E_n) + \epsilon \leq \epsilon$ for all $n \geq n_0$, i. e. $\lim\limits_{n \to \infty} w(A_n) = 0.\square$

**Proposition IV.1.9** *Let the centered Gaussian measure $w$ from Definition IV.1.6 be countably additive. Then, $w(\overline{B_r(a)}) > 0$ for any $a \in H$ and $r > 0$.*

The <u>Proof</u> of this statement is taken from the paper [109]. Again, $\overline{B_r(a)} = \bigcap\limits_{n=1}^{\infty} M_n$ where $M_n = \{x \in H : [(x - a, e_1)_H]^2 + ... + [(x - a, e_n)_H]^2 \leq r^2\}$. Therefore,

$$w(\overline{B_r(a)}) = \lim\limits_{n \to \infty} w(M_n). \text{ Fix } n_0 > 0 \text{ such that } \sum\limits_{k=n_0+1}^{+\infty} \lambda_k < \frac{r^2}{16} \text{ and } \left(\sum\limits_{k=n_0+1}^{+\infty} a_k^2\right)^{\frac{1}{2}} < \frac{r}{4}$$

where $a_k = (a, e_k)_H$. Taking $n \geq n_0 + 1$ we get

$$w(M_n) = (2\pi)^{-\frac{n}{2}} \prod\limits_{k=1}^{n} \lambda_k^{-\frac{1}{2}} \int\limits_{F_n} e^{-\frac{1}{2}\sum\limits_{k=1}^{n} \lambda_k^{-1} x_k^2} dx_1...dx_n \geq$$

$$\geq C(2\pi)^{-\frac{n-n_0}{2}} \prod\limits_{k=n_0+1}^{n} \lambda_k^{-\frac{1}{2}} \int\limits_{F_n^1} e^{-\frac{1}{2}\sum\limits_{k=n_0+1}^{n} \lambda_k^{-1} z_k^2} dz_{n_0+1}...dz_n,$$

where $C$ is a positive constant independent of $n$, $F_n = \{y = (y_1, ..., y_n) \in R^n : (y_1 - a_1)^2 + ... + (y_n - a_n)^2 \leq r^2\}$ and $F_n^1 = \{y = (y_{n_0+1}, ..., y_n) \in R^{n-n_0} : (y_{n_0+1} - a_{n_0+1})^2 + ... + (y_n - a_n)^2 \leq \frac{r^2}{4}\}$ (this is valid because $\{y \in R^n : (y_1 - a_1)^2 + ... + (y_{n_0} - a_{n_0})^2 \leq \frac{r^2}{4}\} \cap \{y \in R^n : (y_{n_0+1} - a_{n_0+1})^2 + ... + (y_n - a_n)^2 \leq \frac{r^2}{4}\} \subset F_n$).

Further, since according to the choice of $n_0$ the embedding

$$\{z = (z_{n_0+1}, ..., z_n) : z_{n_0+1}^2 + ... + z_n^2 \leq \frac{r^2}{16}\} \subset F_n^1$$

is valid, we get:

$$w(M_n) \geq C(2\pi)^{-\frac{n-n_0}{2}} \prod\limits_{k=n_0+1}^{n} \lambda_k^{-\frac{1}{2}} \int\limits_{z_{n_0+1}^2 + ... + z_n^2 \leq \frac{r^2}{16}} e^{-\frac{1}{2}\sum\limits_{k=n_0+1}^{n} \lambda_k^{-1} z_k^2} dz_{n_0+1}...dz_n.$$

Now we use the first inequality from Lemma IV.1.2. According to this result and the above inequality we have:

$$w(M_n) \geq C(1 - \frac{16}{r^2} \sum\limits_{k=n_0+1}^{+\infty} \lambda_k) \geq C_1,$$

where $C_1$ is a positive constant. □

Let $w$ be the centered Gaussian measure from Definition IV.1.6. We define a sequence of Borel measures (finite-dimensional Gaussian measures) $\{w_n\}_{n=1,2,3,...}$ as follows. Let us take and fix an arbitrary integer $n > 0$ and consider the set $\mathcal{M}_n$ of all subsets of the space $H$ of the kind (IV.1.3) with our fixed $n$ and arbitrary Borel sets $F \subset R^n$. Clearly, $\mathcal{M}_n$ is a sigma-algebra and, setting

$$ w_n(M) = (2\pi)^{-\frac{n}{2}} \prod_{i=1}^{n} \lambda_i^{-\frac{1}{2}} \int_F e^{-\frac{1}{2}\sum_{i=1}^{n} \lambda_i^{-1} x_i^2} dx_1...dx_n, $$

we obviously get a countably additive measure $w_n$ defined on the sigma-algebra $\mathcal{M}_n$. Repeating this procedure for all integer $n > 0$, we get our sequence of finite-dimensional Gaussian measures $\{w_n\}_{n=1,2,3,...}$.

Now we show that each measure $w_n$ can be naturally extended onto the whole Borel sigma-algebra $\mathcal{M}$ in $H$ by the rule: $w_n(A) = w_n(A \cap H_n)$, $A \in \mathcal{M}$ (we recall that $H_n = \text{span} \{e_1, ..., e_n\}$). To prove this, it suffices to show that $A \cap H_n$ is a Borel subset of $H_n$ if $A \in \mathcal{M}$. Suppose the opposite. Then, there exists $A \in \mathcal{M}$ such that $A \cap H_n \notin \mathcal{M}_n$. Clearly, $\mathcal{M}_n^1 = \{C \subset H : C = A \cap H_n \text{ for some } A \in \mathcal{M}\}$ is a sigma-algebra and $\mathcal{M}_n^1 \subset \mathcal{M}_n$, $\mathcal{M}_n^1 \neq \mathcal{M}_n$ by the supposition. Consider the set $\mathcal{M}_1$ of all Borel subsets $A$ of $H$ such that $A \cap H_n \in \mathcal{M}_n^1$. Then, one can easily verify that $\mathcal{M}^1$ is a sigma-algebra in $H$ contained in $\mathcal{M}$ and not coinciding with $\mathcal{M}$. But then, since obviously $\mathcal{M}_1$ contains all open and closed subsets of $H$, we get a contradiction because by definition the Borel sigma-algebra is the minimal sigma-algebra containing all open and closed sets. So, $w_n$ can be considered as Borel measures in $H$.

Now, let $\nu, \nu_n$ $(n = 1, 2, 3, ...)$ be nonnegative Borel measures in a complete separable metric space $M$ such that $\nu(M) = \nu_n(M) = 1$, $n = 1, 2, 3, ....$ We recall that the sequence $\{\nu_n\}_{n=1,2,3,...}$ is called *weakly converging* to the measure $\nu$ in $M$ if and only if

$$ \lim_{n\to\infty} \int_M \varphi(x)\nu_n(dx) = \int_M \varphi(x)\nu(dx) $$

for an arbitrary real-valued bounded continuous functional $\varphi$ in $M$.

Lemma IV.1.10 *Let the measure $w$ from Definition IV.1.6 be countably additive. Then, the sequence of measures $\{w_n\}$ weakly converges to the measure $w$ in $H$ as $n \to \infty$.*

Proof. First of all, one can prove as in the proof of Theorem IV.1.8 that for any $\epsilon > 0$ there exists a compact set $K_\epsilon \subset H$ such that $w(K_\epsilon) > 1 - \epsilon$ and $w_n(K_\epsilon) > 1 - \epsilon$

for all integer $n > 0$. Further, let us take an arbitrary real-valued continuous bounded functional $\varphi$ defined in $H$ and prove that

$$\lim_{n\to\infty} \int_H \varphi(x)w_n(dx) = \int_H \varphi(x)w(dx). \tag{IV.1.4}$$

Take also an arbitrary $\epsilon > 0$. Then, one can easily verify that there exists $\delta = \delta(\epsilon) > 0$ such that

$$|\varphi(x) - \varphi(y)| < \epsilon \tag{IV.1.5}$$

for any $x \in K_\epsilon$ and $y \in H$ satisfying $|x - y|_H < \delta$.

Let $K_n = K_\epsilon \cap H_n$, $n = 1, 2, 3, \dots$. Obviously, for any integer $n > 0$

$$\left| \int_H \varphi(x)w_n(dx) - \int_{K_n} \varphi(x)w_n(dx) \right| < \epsilon M,$$

where $M = \sup_{x \in H} |\varphi(x)|$, so that we have

$$\liminf_{n\to\infty} \left| \int_H \varphi(x)w_n(dx) - \int_{K_n} \varphi(x)w_n(dx) \right| \leq M\epsilon. \tag{IV.1.6}$$

Let us now show the existence of $C > 0$, depending only on $\varphi$ and independent of $\epsilon > 0$, such that

$$\liminf_{n\to\infty} \left| \int_H \varphi(x)w(dx) - \int_{K_n} \varphi(x)w_n(dx) \right| \leq C\epsilon. \tag{IV.1.7}$$

Clearly, in view of the arbitrariness of $\epsilon > 0$, relations (IV.1.6) and (IV.1.7) together yield (IV.1.4), i. e. the statement of the lemma. Let $K_{n,\epsilon} =$

$$= \left\{ y \in H : y = y_1 + y_2, \ y_1 \in H_n, \ y_2^\perp \in H_n, \ |y_2|_H < \frac{\delta(\epsilon)}{2}, \ \text{dist}\,(y_1, K_n) < \frac{\delta(\epsilon)}{2} \right\}.$$

Then, $K_\epsilon \subset K_{n,\epsilon}$ for all sufficiently large numbers $n$. Hence,

$$\left| \int_H \varphi(x)w(dx) - \int_{K_{n,\epsilon}} \varphi(x)w(dx) \right| < M\epsilon$$

for all sufficiently large numbers $n$. Further, it is clear that the measure $w$ is the direct product $w_n \otimes w_n^\perp$ of the measure $w_n$ in $H_n$ and the Gaussian measure $w_n^\perp$ defined in the Hilbert space $H_n^\perp$, which is the orthogonal complement in $H$ of the subspace $H_n$, by the rule: for a cylindrical set

$$M = \{ x \in H_n^\perp : [(x, e_{n+1})_H, \dots, (x, e_{n+k})_H] \in F \},$$

where $F \subset R^k$ is a Borel set,

$$w_n^{\perp}(M) = (2\pi)^{-\frac{k}{2}} \prod_{i=n+1}^{n+k} \lambda_i^{-\frac{1}{2}} \int_F e^{-\frac{1}{2}\sum_{i=n+1}^{n+k} \lambda_i^{-1} x_i^2} \, dx_{n+1}...dx_{n+k}.$$

Here, in particular, $w_n(H_n) = w_n^{\perp}(H_n^{\perp}) = 1$. For $x_n \in H_n$ let also $K_{n,\epsilon}^{\perp}(x_n) = K_{n,\epsilon} \cap \{x \in H : x = x_n + y, \ y \in H_n^{\perp}\}$. Then, by (IV.1.5)

$$\int_{K_{n,\epsilon}} \varphi(x) w(dx) = \int_{x_n \in K_{n,\epsilon}} w_n(dx_n) \int_{x_n^{\perp} \in K_{n,\epsilon}^{\perp}(x_n)} \varphi(x_n + x_n^{\perp}) w_n^{\perp}(dx_n^{\perp}) =$$

$$= \alpha(n,\epsilon) + \int_{K_{n,\epsilon}} \varphi(x_n) w_n(dx_n),$$

where $|\alpha(n,\epsilon)| \leq C_2 \epsilon$ for some $C_2 > 0$ independent of $\epsilon$ and $n$.$\square$

Now, let $\Phi(x)$ be a real-valued nonnegative continuous functional in $H$ bounded on bounded subsets of $H$. Consider the quantities

$$\mu_n(\Omega) = \int_{\Omega} \Phi(x) w_n(dx) \quad \text{and} \quad \mu(\Omega) = \int_{\Omega} \Phi(x) w(dx),$$

where $\Omega$ is an arbitrary bounded Borel subset of $H$.

<u>Lemma IV.1.11</u> $\liminf_{n \to \infty} \mu_n(\Omega) \geq \mu(\Omega)$ *for an arbitrary open bounded set* $\Omega \subset H$. $\limsup_{n \to \infty} \mu_n(K) \leq \mu(K)$ *for an arbitrary closed bounded set* $K \subset H$.

<u>Proof</u> is standard. Here we present its variant taken from the paper [109]. Let us prove, for example, the first statement (the second can be proved by analogy). Fix an arbitrary $\epsilon > 0$ and take a real-valued functional $\psi_\epsilon(u)$ continuous in $H$ and satisfying the following properties:

1. $0 \leq \psi_\epsilon(u) \leq 1$ for all $u \in X$;
2. $\psi_\epsilon(u) = 0$ for any $u \notin \Omega$;
3. $\psi_\epsilon(u) = 1$ if $u \in \Omega$ and dist$(u, \partial\Omega) \geq \epsilon$.

Then, according to Lemma IV.1.10, we have:

$$\liminf_{n \to \infty} \mu_n(\Omega) \geq \lim_{n \to \infty} \int_{\Omega} \psi_\epsilon(u)\mu_n(du) = \lim_{n \to \infty} \int_{\Omega} \psi_\epsilon(u)\Phi(u)w_n(du) =$$

$$= \int_{\Omega} \psi_\epsilon(u)\Phi(u)w(du) = \int_{\Omega} \psi_\epsilon(u)d\mu(u).$$

Taking here the limit over a sequence $\epsilon_r \to +0$ and an arbitrary sequence of functionals $\psi_{\epsilon_r}(u)$ satisfying the above properties with $\epsilon = \epsilon_r$, we get $\liminf_{n \to \infty} \mu_n(\Omega) \geq \mu(\Omega)$.$\square$

## 4.2   An invariant measure for the NLSE

In this section, we shall construct an invariant measure for the NLSE; we follow the paper [107]. Let $A > 0$, the space $L_2(0, A)$ be real and let $X$ be the direct product $L_2(0, A) \otimes L_2(0, A)$ of two samples of the space $L_2(0, A)$ equipped with the scalar product

$$(w_1, w_2)_X = (u_1, u_2)_{L_2(0,A)} + (v_1, v_2)_{L_2(0,A)},$$

where $w_i = (u_i, v_i)$, $u_i, v_i \in L_2(0, A)$, $i = 1, 2$, and the corresponding norm $\|u\|_X = (u, u)_X^{\frac{1}{2}}$. Let us consider the following problem for the NLSE written in the real form as a system of equations for real and imaginary parts of the unknown function:

$$u_t^1 + u_{xx}^2 + f(x, (u^1)^2 + (u^2)^2)u^2 = 0, \quad x \in (0, A), \quad t \in R, \tag{IV.2.1}$$

$$u_t^2 - u_{xx}^1 - f(x, (u^1)^2 + (u^2)^2)u^1 = 0, \quad x \in (0, A), \quad t \in R, \tag{IV.2.2}$$

$$u^i(0, t) = u^i(A, t) = 0, \quad t \in R, \tag{IV.2.3}$$

$$u^i(x, t_0) = u_0^i \in L_2(0, A). \tag{IV.2.4}$$

Formally the problem (IV.2.1)-(IV.2.4) is equivalent to the integral equation

$$u(t) = A(t - t_0)u_0 + \int_{t_0}^{t} B(t - s)[f(\cdot, |u(s)|^2)u(s)]ds, \tag{IV.2.5}$$

where $u(t) = (u^1(t), u^2(t))$ is the unknown function with values in a functional space (to be interpreted as $X$), $u_0 = (u_0^1, u_0^2)$,

$$A(t) = \begin{pmatrix} \cos(tD) & \sin(tD) \\ -\sin(tD) & \cos(tD) \end{pmatrix} \quad \text{and} \quad B(t) = \begin{pmatrix} \sin(tD) & -\cos(tD) \\ \cos(tD) & \sin(tD) \end{pmatrix}.$$

In this section, our hypothesis on the function $f$ is the following.

(f) *Let $f(x, s)$ be a real-valued continuous function of $(x, s) \in [0, A] \times [0, +\infty)$ possessing a continuous partial derivative $f_s'(x, s)$ and let there exist $C > 0$ such that*

$$|f(x, s)| + |f_s'(x, s)| \le C$$

*for all $(x, s) \in [0, A] \times [0, +\infty)$.*

One may define $X$-solutions of the problem (IV.2.1)-(IV.2.4) by analogy with $H^1$-solutions of the NLSE from Definition I.2.2.

Proposition IV.2.1 *Under the hypothesis (f) an arbitrary solution $u(\cdot, t) \in C(I; X) \cap C^1(I; H^{-2}(0, A) \times H^{-2}(0, A))$ of the problem (IV.2.1)-(IV.2.4) satisfies equation (IV.2.5), and conversely, any solution $u(\cdot, t) \in C(I; X)$ of equation (IV.2.5) is a solution of the problem (IV.2.1)-(IV.2.4) of the class $C(I; X) \cap C^1(I; H^{-2}(0, A) \times H^{-2}(0, A))$ (here $I = [t_0 - T_1, t_0 + T_2]$ where $T_1, T_2 > 0$).*

Proof can be made by analogy with the proof of Proposition I.2.3□

Now we can present the main result of this section.

Theorem IV.2.2 *Under the hypothesis (f)*

*(a) for any $u_0 \in X$ there exists a unique global $X$-solution $u(\cdot, t)$ of the problem (IV.2.1)-(IV.2.4);*

*(b) let $h((u_0^1, u_0^2), t)$ be the function transforming any pair $(u_0^1, u_0^2) \in X$ and $t \in R$ into the $X$-solution $(u^1(\cdot, t + t_0), u^2(\cdot, t + t_0))$ of the problem (IV.2.1)-(IV.2.4) taken at the moment of time $t + t_0$. Then, the function $h$ is a dynamical system with the phase space $X$;*

*(c) $\frac{d}{dt}\|u(\cdot, t)\|_X^2 = 0$ for an arbitrary above $X$-solution of the problem (IV.2.1)-(IV.2.4);*

*(d) let $w$ be the centered Gaussian measure with the correlation operator $S = \begin{pmatrix} D^{-1} & 0 \\ 0 & D^{-1} \end{pmatrix}$ in the space $X$. Since $S$ is an operator of trace class, the measure $w$ is countably additive. Let also $\Phi(u^1, u^2) = \int_0^A F(x, (u^1)^2 + (u^2)^2)dx$ where $F(s) = \frac{1}{2}\int_0^s f(x, p)dp$ (the functional $\Phi$ is obviously real-valued and continuous in the space $X$ bounded on bounded subsets of $X$). Then, the Borel measure in $X$*

$$\mu(\Omega) = \int_\Omega e^{\Phi(u^1, u^2)} w(du^1 \, du^2)$$

*(here $\Omega$ is an arbitrary Borel subset of $X$) is well-defined in $X$ and it is an invariant measure for the dynamical system $h$ with the phase space $X$.*

Proof of Theorem IV.2.2. The map from the right-hand side of (IV.2.5) transforms the space $C(I; X)$ into itself and is a contraction for sufficiently small values of $T > 0$ depending only on $\|u_0\|_X$ (here $I = [t_0 - T, t_0 + T]$). Therefore, equation (IV.2.5) has a unique local solution of the class $C(I; X)$. Further, let $\{\lambda_n, e_n\}_{n=0,1,2,...}$ be eigenvalues and the corresponding eigenfunctions of the operator $D$ and let $X_n = \text{span}(e_0, ..., e_n)$ and $P_n$ be the orthogonal projector in the space $L_2(0, A)$ onto the subspace $X_n$. Let also $X^n = X_n \otimes X_n$. Consider the following problem approximating the problem (IV.2.1)-(IV.2.4):

$$u^1_{n,t} + u^2_{n,xx} + P_n[f(x, (u^1_n)^2 + (u^2_n)^2)u^2_n] = 0, \quad t \in R, \qquad (IV.2.6)$$

$$u^2_{n,t} - u^1_{n,xx} - P_n[f(x, (u^1_n)^2 + (u^2_n)^2)u^1_n] = 0, \quad t \in R, \qquad (IV.2.7)$$

$$u^1_n(x, t_0) = P_n u^1_0(x), \quad u^2_n(x, t_0) = P_n u^2_0(x). \qquad (IV.2.8)$$

Let $P^n = \begin{pmatrix} P_n & 0 \\ 0 & P_n \end{pmatrix}$ be the orthogonal projector onto the subspace $X^n = X_n \otimes X_n$
in the space $X$. Let also $g_1 = (e_0, 0), g_2 = (0, e_0), ..., g_{2n+1} = (e_n, 0), g_{2n+2} = (0, e_n), ....$
Then, the system $\{g_n\}_{n=1,2,3,...}$ is an orthonormal basis in the space $X$ consisting of
the eigenelements of the operator $S$. Clearly, for any positive integer $n$, the problem
(IV.2.6)-(IV.2.8) has a unique local solution $u_n(x, t) = (u^1_n(x, t), u^2_n(x, t)) \in C(I; X^n)$
(as it is well known, in a finite-dimensional linear space any two norms are equivalent,
and we mean that the space $X^n$ is equipped with the norm of the space $X$). In
addition, the direct verification shows that $\frac{d}{dt}\|u_n(\cdot, t)\|^2_X = 0$ for these solutions.
Therefore, for any $n = 1, 2, 3, ...$ and for any $u_0(\cdot) = (u^1_0(\cdot), u^2_0(\cdot)) \in X$ the problem
(IV.2.6)-(IV.2.8) has a unique global solution $u_n(\cdot, t) \in C(R; X^n)$.

Further, it is clear that the above solutions $u_n(\cdot, t)$ of the problem (IV.2.6)-
(IV.2.8) satisfy the equations ($n = 1, 2, 3, ...$):

$$u_n(\cdot, t) = A(t - t_0)P^n u_0 + \int_{t_0}^t B(t - s)P^n[f(\cdot, |u_n(\cdot, s)|^2)u_n(\cdot, s)]ds. \qquad (IV.2.9)$$

Hence, from (IV.2.5) and (IV.2.9) one has for those values of $t$ for which the solution
$u(\cdot, t)$ exists:

$$\|u(\cdot, t) - u_n(\cdot, t)\|_X \le C_1\|u_0 - P^n u_0\|_X + C_2 \int_{t_0}^t \|u_n(\cdot, s) - u(\cdot, s)\|_X ds +$$

$$+C_3 \int_{t_0}^t \|u(\cdot, s) - P^n u(\cdot, s)\|_X ds. \qquad (IV.2:10)$$

Here the constants $C_1, C_2, C_3$ do not depend on the initial value $u_0$, $t_0$ and $t$. Let
the solution $u(\cdot, t)$ exist for $t \in [t_0, t_0 + T]$ where $T > 0$. Then, the third term in the
right-hand side of this inequality obviously tends to zero as $n \to +\infty$ uniformly with
respect to $t \in [t_0, t_0 + T]$, therefore we get from inequality (IV.2.10) by the Gronwell's
lemma that

$$\lim_{n \to \infty} \max_{t \in [t_0, t_0+T]} \|u(\cdot, t) - u_n(\cdot, t)\|_X = 0.$$

By analogy, if the solution $u(\cdot, t)$ exists on a segment $[t_0 - T, t_0]$, $T > 0$, then

$$\lim_{n \to \infty} \max_{t \in [t_0-T, t_0]} \|u(\cdot, t) - u_n(\cdot, t)\|_X = 0.$$

Hence,

$$\lim_{n\to\infty} \max_{t\in I} ||u(\cdot,t) - u_n(\cdot,t)||_X = 0 \qquad (IV.2.11)$$

for all segments $I = [t_0 - T_1, t_0 + T_2]$ of the existence of the solution $u(\cdot,t)$. This fact implies, in particular, the statement (c) of Theorem IV.2.2 and, hence, the global solvability of equation (IV.2.5) for any $u_0 \in X$.

Further, it is easy to verify that, if a function $u(\cdot,t) \in C(R;X)$ satisfies equation (IV.2.5), then for any fixed $t \in R$ this function is a solution of the following equation:

$$u(\cdot,\tau) = A(\tau - t)u(\cdot,t) + \int_t^\tau B(\tau - s)[f(\cdot,|u(\cdot,s)|^2)u(\cdot,s)]ds, \quad \tau \in R,$$

which, as earlier, for any fixed $t$ has a unique global $X$-solution. Therefore, for any fixed $t$ the map $u_0 \to u(\cdot,t)$ is one-to-one from $X$ into $X$. The continuity of the transformation $u_0 \to u(\cdot,t)$ as a map from $X$ into $X$ follows from the estimate $(t > t_0)$

$$||u(\cdot,t) - v(\cdot,t)||_X \le C_1||u(\cdot,t_0) - v(\cdot,t_0)||_X + C_2 \int_{t_0}^t ||u(\cdot,s) - v(\cdot,s)||_X ds,$$

where $u(\cdot,t)$ and $v(\cdot,t)$ are two arbitrary solutions of equation (IV.2.5), and a similar estimate for $t < t_0$. Thus, the statements (a),(b) and (c) of Theorem IV.2.2 are proved.

<u>Lemma IV.2.3</u> *For any $\epsilon > 0$ and $T > 0$ there exists $\delta > 0$ such that*

$$\max_{t\in[t_0-T,t_0+T]} ||u_n(\cdot,t) - v_n(\cdot,t)||_X < \epsilon$$

*for all numbers $n = 1,2,3,...$ and for any two solutions $u_n(\cdot,t)$ and $v_n(\cdot,t)$ of the problem (IV.2.6)-(IV.2.8), taken with the same value $n$, satisfying the condition*

$$||u_n(\cdot,t_0) - v_n(\cdot,t_0)||_X < \delta$$

*(here $u_n(\cdot,t_0) = P^n u_0$ and $v_n(\cdot,t_0) = P^n v_0$).*
<u>Proof</u> follows from the estimate $(t > t_0)$

$$||u_n(\cdot,t) - v_n(\cdot,t)||_X \le C_1||u_n(\cdot,t_0) - v_n(\cdot,t_0)||_X + C_2 \int_{t_0}^t ||u_n(\cdot,s) - v_n(\cdot,s)||_X ds,$$

that results from equation (IV.2.9), and an analogous estimate for $t < t_0$.□

By $h_n(u_0,t)$ we denote the function mapping any $u_0 \in X$ and $t \in R$ into $u_n(\cdot,t+t_0)$ where $u_n(\cdot,t)$ is the solution of the problem (IV.2.6)-(IV.2.8). It is clear that the function $h_n$ is a dynamical system with the phase space $X^n$. For each $n = 1, 2, 3, \ldots$ let us consider in the space $X^n$ the centered Gaussian measure $w_n$ with the correlation operator $S$. Since $S = S^* > 0$ in $X^n$, the measure $w_n$ is well-defined in $X^n$. According to the result from Section 4.1, measures $w_n$ can be considered as Borel measures in $X$. Also, since $\Phi(u) = \Phi(u^1, u^2)$ is a continuous functional in $X^n$, the following Borel measures

$$\mu_n(\Omega) = \int\limits_\Omega e^{\Phi(u)} dw_n(u)$$

(where $\Omega \subset X$ is an arbitrary Borel set) are well defined.

**Lemma IV.2.4** $\mu_n$ *is an invariant measure for the dynamical system* $h_n$ *with the phase space* $X^n$.

Proof. Let us rewrite the system (IV.2.6)-(IV.2.8) for the coefficients $a_i$, $b_i$ where $u_n^1 = \sum\limits_{i=0}^{n} a_i(t) e_i$ and $u_n^2 = \sum\limits_{i=0}^{n} b_i(t) e_i$. Then, we get

$$\dot{a}(t) = \nabla_b E_n(a, b), \tag{IV.2.12}$$

$$\dot{b}(t) = -\nabla_a E_n(a, b), \tag{IV.2.13}$$

$$a_i(t_0) = (u_n^1(\cdot, t_0), e_i)_{L_2(0,A)}, \quad b_i(t_0) = (u_n^2(\cdot, t_0), e_i)_{L_2(0,A)}, \quad (i = 1, 2, \ldots, n), \tag{IV.2.14}$$

where $a(t) = (a_0(t), \ldots, a_n(t))$, $b(t) = (b_0(t), \ldots, b_n(t))$, $u_n = (u_n^1, u_n^2)$ and $E_n(a, b) = E(u_n) = \int\limits_0^A \left\{ \frac{1}{2} \left[ (u_{n,x}^1)^2 + (u_{n,x}^2)^2 \right] - F(x, (u_n^1)^2 + (u_n^2)^2) \right\} dx$. Then, according to Theorem IV.1.3, the dynamical system with the phase space $R^{2(n+1)}$ generated by the system (IV.2.12)-(IV.2.14) possesses a Borel invariant measure $\mu_n'$:

$$\mu_n'(A) = (2\pi)^{-(n+1)} \prod\limits_{i=0}^{n} \lambda_i \int\limits_A e^{E_n(a,b)} da\, db,$$

where $A \subset R^{2(n+1)}$ is an arbitrary Borel set. Further, according to Proposition IV.1.1, there is a natural one-to-one correspondence between Borel subsets $A$ of the space $R^{2(n+1)}$ and Borel subsets $\Omega$ of the space $X^n$ defined by the rule: a Borel set $A \subset R^{2(n+1)}$ corresponds to a Borel set $\Omega \subset X^n$ if an element $u_n = (u_n^1, u_n^2)$ belongs to $\Omega$ when and only when $(a, b) \in A$ where $u_n^1 = \sum\limits_{i=0}^{n} a_i e_i$, $u_n^2 = \sum\limits_{i=0}^{n} b_i e_i$ and $a = (a_0, \ldots, a_n)$, $b = (b_0, \ldots, b_n)$. In addition, if two sets $A \subset R^{2(n+1)}$ and $\Omega \subset X^n$ correspond to each other in this sense, then $\mu_n(\Omega) = \mu_n'(A)$ by the definition of the measure $\mu_n$. These arguments easily imply the statement of Lemma IV.2.4. $\square$

According to Lemma IV.1.10, the sequence of Borel measures $w_n$ weakly converges to the measure $w$ as $n \to \infty$.

<u>Lemma IV.2.5</u> $\mu(\Omega) = \mu(h(\Omega, t))$ *for any open bounded set* $\Omega \subset X$ *and for any* $t \in R$.

<u>Proof.</u> Fix an arbitrary $t \in R$ and let $\Omega \subset X$ be an arbitrary open bounded set. Then, according to the proved statements (a),(b),(c) of Theorem IV.2.1, $h(\Omega, t)$ is open and bounded, too. Let us fix an arbitrary $\epsilon > 0$. Then, since the functional $\Phi$ is bounded on bounded subsets of the space $X$ and in view of results from Section 4.1, there exists a compact set $K \subset \Omega$ such that $\mu(\Omega \setminus K) < \epsilon$. Further, according to the proved statements (a) and (b) of Theorem IV.2.2, $K_1 = h(K, t)$ is a compact set, too, and $K_1 \subset \Omega_1 = h(\Omega, t)$.

For any $A \subset X$, let $\partial A$ be the boundary of the set $A$ and let

$$\alpha = \min\{\operatorname{dist}(K, \partial\Omega); \operatorname{dist}(K_1, \partial\Omega_1)\}$$

(where again $\operatorname{dist}(A, B) = \inf_{x \in A, \, y \in B} ||x - y||_X$). Then, obviously, $\alpha > 0$. According to Lemma IV.2.3, for any $x \in K$ there exists $\delta > 0$ such that for any $u, v \in B_\delta(x) = \{y \in X| \; ||y - x||_X < \delta\}$ and for any $n = 1, 2, 3, \ldots$ one has $||h_n(u, t) - h_n(v, t)||_X < \frac{\alpha}{3}$ and $B_\delta(x) \subset \Omega$. Let $B_{\delta_1}(x_1), \ldots, B_{\delta_l}(x_l)$ be a finite covering of the compact set $K$ by these balls and let $B = \bigcup_{i=1}^{l} B_{\delta_i}(x_i)$. Then, by (IV.2.11) and by construction, we get the existence of a number $n_0 > 0$ such that $\operatorname{dist}(h_n(u, t); \partial\Omega_1) \geq \frac{\alpha}{2}$ for all $u \in B$ and for all $n \geq n_0$. Further, we get by Lemmas IV.1.11 and IV.2.4

$$\mu(\Omega) \leq \mu(B) + \epsilon \leq \liminf_{n \to \infty} \mu_n(B) + \epsilon \leq \liminf_{n \to \infty} \mu_n(h_n(B, t)) + \epsilon \leq \mu(\Omega_1) + \epsilon$$

(because $\mu_n(B) = \mu_n(B \cap X^n) = \mu_n(h_n(B \cap X^n, t))$ and $h_n(B \cap X^n, t) \subset h_n(B, t)$). Hence, due to the arbitrariness of $\epsilon > 0$, we have $\mu(\Omega) \leq \mu(\Omega_1)$. By analogy $\mu(\Omega) \geq \mu(\Omega_1)$. Thus, $\mu(\Omega) = \mu(\Omega_1)$, and Lemma IV.2.5 is proved.□

According to the proved statement (c) of Theorem IV.2.2, we have for an arbitrary open set $\Omega \subset X$:

$$\mu(\Omega) = \lim_{R \to +\infty} \mu(B_R(0) \cap \Omega) = \lim_{R \to +\infty} \mu(B_R(0) \cap h(\Omega, t)) = \mu(h(\Omega, t)).$$

For an arbitrary Borel set $A \subset X$ we obtain the equality $\mu(A) = \mu(h(A, t))$ by the approximation of the set $A$ by open sets containing $A$. Thus, Theorem IV.2.2 is proved.□

<u>Remark IV.2.6</u> Generally, the invariant measure $\mu$ given by Theorem IV.2.2 can be unbounded, i. e. it may happen that $\mu(X) = +\infty$. However, according to the

statement (c) of this theorem, any ball $B_R(0)$ with $R > 0$ is an invariant set of the dynamical system $h$. Therefore, we can choose any of these balls for a new phase space of the dynamical system $h$. Since the functional $\Phi$ is bounded on any of such a ball, we get that $0 < \mu(B_R(0)) < +\infty$ for any $R > 0$. By analogy, in view of Proposition IV.1.9, $0 < \mu(B_r(a)) < +\infty$ for any $a \in X$ and $r > 0$. As a corollary, we obtain that almost all points of the space $X$ (in the sense of the measure $w$) are stable according to Poisson and the set of all these points is dense in the space $X$.

<u>Remark IV.2.7</u> The hypothesis (f) of Theorem IV.2.2 is valid, for example, for two physical nonlinearities: $f(x,s) = \frac{\alpha s}{1+s}$ and $f(x,s) = e^{-\alpha s}$ where $\alpha \geq 0$ in the second case.

## 4.3  An infinite series of invariant measures for the KdVE

In this section, we consider the following Cauchy problem periodic with respect to the spatial variable for the standard KdVE with $f(u) = u$:

$$u_t + u u_x + u_{xxx} = 0, \quad x, t \in R, \tag{IV.3.1}$$

$$u(x + A, t) = u(x, t), \quad x, t \in R, \tag{IV.3.2}$$

$$u(x, t_0) = u_0(x), \tag{IV.3.3}$$

where $u_0(x + A) = u_0(x)$, $x \in R$, and $A > 0$ is fixed. Here we follow the paper [111]. We shall construct an infinite series of invariant measures associated with conservation laws $E_n$, $n \geq 3$, given by Theorem I.1.5 for dynamical systems generated by the problem (IV.3.1)-(IV.3.3) on suitable phase spaces. Throughout this section we shall denote the norm in the space $H^n_{\text{per}}(A)$ by $\| \cdot \|_n$ for simplicity. Our first related result is the following.

<u>Theorem IV.3.1</u> *Let integer $n \geq 2$, $A > 0$, $T > 0$, $t \in R$ and $t_0 \in R$ be arbitrary. Then, the function $h^n$ from $H^n_{\text{per}}(A)$ into $H^n_{\text{per}}(A)$ defined for any $u_0 \in H^n_{\text{per}}(A)$ by the rule: $h^n(u_0, t) = u(\cdot, t + t_0)$, where $u(\cdot, t)$ is the $H^n_{\text{per}}(A)$-solution of the problem (IV.3.1)-(IV.3.3) given by Theorem I.1.5, is a dynamical system with the phase space $H^n_{\text{per}}(A)$. The functionals $E_0, ..., E_{n-1}$ from Theorem I.1.5 are conservation laws for this dynamical system.*

Let $n \geq 2$ be integer and $w^n$ be the centered Gaussian measure in $H^{n-1}_{\text{per}}(A)$ with the correlation operator $S = (I + D^{n-1})(D^n + I)^{-1}$. Since clearly $S$ is an operator of trace class in $H_{per}^{n-1}(A)$, the measure $w^n$ is countably additive. Let also for $u \in H_{per}^n(A)$

$$J_n(u) = E_n(u) - \frac{1}{2} \int_0^A \{[u_x^{(n)}]^2 + u^2\}dx = E_n(u) - \frac{1}{2}(S^{-1}u, u)_{H_{per}^n(A)} =$$

$$= \int_0^A \{c_n u[u_x^{(n-1)}]^2 - \frac{1}{2}u^2 - q_n(u, ..., u_x^{(n-2)})\}dx.$$

For an arbitrary Borel $\Omega \subset H_{per}^{n-1}(A)$ we set

$$\mu^n(\Omega) = \int_\Omega e^{-J_n(u)}dw^n(u),$$

where $\Omega \subset H_{per}^{n-1}(A)$ is a Borel set. The main result of this section is the following.

Theorem IV.3.2 *For any integer* $n \geq 3$ $\mu^n$ *is a well-defined nonnegative Borel measure and it is an invariant measure for the dynamical system* $h^{n-1}$ *given by Theorem IV.3.1. For any sufficiently large* $d > 0$ $0 < \mu^n(R_d) < +\infty$, *where*

$$R_d = \{u \in H_{per}^{n-1}(A): E_k(u) \leq d, \quad k = 0, 1, ..., n-1\}.$$

*Since due to Theorem IV.3.1* $R_d$ *is an invariant set of the dynamical system* $h^{n-1}$, *it can be taken for a new phase space, therefore the Poincaré Recurrence Theorem is applicable. So, almost all points of the space* $H_{per}^{n-1}(A)$ *(in the sense of the measure* $w^n$*) are stable according to Poisson and, in addition, all points stable according to Poisson form a dense set in* $H_{per}^{n-1}(A)$.

Theorem IV.3.1 immediately follows from Theorem I.1.5 (and the proof of the latter). Now we turn to proving Theorem IV.3.2. Let

$$e_0(x) \equiv \frac{1}{\sqrt{A}}, \quad e_{2k-1}(x) = \sqrt{\frac{2}{A(1 + \lambda_{2k-1}^n)}} \sin \frac{2\pi kx}{A}, \quad e_{2k}(x) = \sqrt{\frac{2}{A(1 + \lambda_{2k}^n)}} \cos \frac{2\pi kx}{A},$$

where $k = 1, 2, 3, ....$ Then, $\{e_m\}_{m=0,1,2,...}$ is an orthonormal basis of the space $H_{per}^n(A)$ consisting of eigenfunctions of the operator $D$ with the corresponding eigenvalues $0 = \lambda_0 < \lambda_1 = \lambda_2 < ... < \lambda_{2k-1} = \lambda_{2k} < ....$ Let $P_m$ be the orthogonal projector in $H_{per}^n(A)$ onto the subspace $L_m = \text{span}\{e_0, ..., e_{2m}\}$ and $P_m^\perp$ be the orthogonal projector in $H_{per}^n(A)$ onto the orthogonal complement $L_m^\perp$ to the subspace $L_m$. Consider the following problem:

$$u_t^m + P_m[u^m u_x^m] + u_{xxx}^m = 0, \quad x \in (0, A), \ t \in R, \qquad (IV.3.4)$$

$$u^m(x, t_0) = P_m u_0(x). \tag{IV.3.5}$$

Clearly, for any $u_0 \in H^n_{\text{per}}(A)$ it has a unique classical local solution $u^m(x, t) \in C^1([t_0 - T, t_0 + T]; L_m)$ for some $T > 0$ (the topology in $L_m$ is generated by the norm of $H^n_{\text{per}}(A)$). Since it can be easily verified that $\frac{d}{dt} E_0(u^m) = \frac{d}{dt} E_1(u^m) = 0$, we have:

$$|u^m(\cdot, t)|_{L_2(0,A)} = |u^m(\cdot, t_0)|_{L_2(0,A)}$$

for all $t$, and, hence, the solution of the problem (IV.3.4),(IV.3.5) for any $u_0 \in H^n_{\text{per}}(A)$ is global (it can be continued onto the entire real line $t \in R$). (We apply here the known fact that in a finite-dimensional linear space any two norms are equivalent). In addition, this solution is obviously infinitely differentiable in $x$ and $t$.

**Proposition IV.3.3** *Let $n \geq 2$ be integer, $u_0 \in H^n_{\text{per}}(A)$ and let a sequence $u_0^{m_k}$ be such that $m_k \to +\infty$ as $k \to \infty$, $u_0^{m_k} \in L_{m_k}$ and $u_0^{m_k} \to u_0$ strongly in $H^n_{\text{per}}(A)$ as $k \to \infty$. Then for any $t \in R$ $u^{m_k}(\cdot, t) \to u(\cdot, t)$ as $k \to \infty$ strongly in $H^n_{\text{per}}(A)$, where $u(\cdot, t)$ is the $H^n_{\text{per}}(A)$-solution of the problem (IV.3.1)-(IV.3.3) and $u^{m_k}(\cdot, t)$ is the $L_{m_k}$-solution of equation (IV.3.4) with $m = m_k$ and with the initial data*

$$u^{m_k}(\cdot, t_0) = u_0^{m_k}.$$

Before proving this statement consider a number of lemmas.

**Lemma IV.3.4** *For any nonnegative integer $n$ and $d > 0$ there exists $R = R(n, d) > 0$ such that if $u \in H^n_{\text{per}}(A)$ and*

$$E_0(u) < d, ..., E_n(u) < d,$$

*then $\|u\|_n < R$.*

Proof. Consider the conservation law ($n \geq 2$):

$$\frac{1}{2}E_0(u) + E_n(u) = \int\limits_0^A \left\{ \frac{1}{2}(D_x^n u)^2 + \frac{1}{2}u^2 + c_n u(D_x^{n-1}u)^2 - q_n(u, ..., D_x^{n-2}u) \right\} dx \geq$$

$$\geq \frac{1}{2}\|u(.)\|_n^2 - \eta_n(\|u(.)\|_{n-1}),$$

where $\eta_n(s)$ is a function continuous and increasing on $[0, +\infty)$. Repeat these estimates for the functionals $\frac{1}{2}E_0(u) + E_2(u), ..., \frac{1}{2}E_0(u) + E_{n-1}(u)$. For the functional $E_1(u)$ we have in view of the known inequality $|u|_{L_p(0,A)} \leq |u|_{L_2(0,A)}^{\frac{1}{2}+\frac{1}{p}}(|D_x u|_{L_2(0,A)} + |u|_{L_2(0,A)})^{\frac{1}{2}-\frac{1}{p}}$, where $p \geq 2$:

$$E_1(u) + \frac{1}{2}E_0(u) \geq \frac{1}{2}\|u\|_1^2 - \eta_1(|u|_{L_2(0,A)})(\|u\|_1^{\frac{1}{2}} + 1).$$

We get step by step from the obtained estimates:

$$||u||_1 \leq C_1(d), ..., ||u||_n \leq C_n(d)$$

for all $t \in R$, and Lemma IV.3.4 is proved.□

<u>Lemma IV.3.5</u> *There exist functions* $\gamma_n(R, s)$, *such that* $\gamma_n(R, 0) \equiv 0$, *defined on* $(R, s) \in [0, +\infty) \otimes [0, +\infty)$, *monotonically nondecreasing in* $s$ *and continuous satisfying the following:*

$$\frac{d}{dt} E_n(u^{m_k}(\cdot, t)) \leq$$

$$\leq \gamma_n(R, \max_{\substack{0 \leq i,j \leq n-1 \\ i+j \neq 2n-2}} |P_{m_k}^{\perp}[D_x^i u^{m_k}(\cdot, t) D_x^j u^{m_k}(\cdot, t)]|_{L_2(0,A)} + ||P_{m_k}^{\perp}(u^{m_k} u_x^{m_k})||_1)$$

*for all* $t \in R$, *all* $n = 3, 4, 5, ...$, *all* $k = 1, 2, 3, ...$, *all* $u_0 \in H_{per}^{n-1}(A)$, *for which* $||u_0||_{n-1} \leq R$, *and all above sequences* $u_0^{m_k}$.

<u>Proof.</u> Let $Lg = -gg_x - g_{xxx}$. For $u_0 \in H_{per}^{n-1}(A)$, since under the substitution of $Lu^{m_k}$ in place of $u_t^{m_k}$ in the integrand from the expression for $\frac{d}{dt} E_n(u^{m_k}(\cdot, t))$ we get zero (see [50]), we have:

$$\frac{d}{dt} E_n(u^{m_k}(\cdot, t)) = \int_0^A \frac{\partial}{\partial t} \left\{ \frac{1}{2}(D_x^n u^{m_k})^2 + c_n u^{m_k}(D_x^{n-1} u^{m_k})^2 - \right.$$

$$\left. -q_n(u^{m_k}, ..., D_x^{n-2} u^{m_k}) \right\} \bigg|_{u_t^{m_k} = P_{m_k}^{\perp}(u^{m_k} u_x^{m_k})} dx = \int_0^A \left\{ (-1)^n D_x^{2n} u^{m_k} P_{m_k}^{\perp}(u^{m_k} u_x^{m_k}) + \right.$$

$$+c_n P_{m_k}^{\perp}(u^{m_k} u_x^{m_k})(D_x^{n-1} u^{m_k})^2 + (-1)^{n-1} 2c_n D_x^{n-1}(u^{m_k} D_x^{n-1} u^{m_k}) \times P_{m_k}^{\perp}(u^{m_k} u_x^{m_k}) +$$

$$\left. +\sum_{i=0}^{n-2} \frac{\partial q_n(u^{m_k}, ..., D_x^{n-2} u^{m_k})}{\partial (D_x^i u^{m_k})} D_x^i P_{m_k}^{\perp}(u^{m_k} u_x^{m_k}) \right\} = \int_0^A \left\{ c_n P_{m_k}^{\perp}(u^{m_k} u_x^{m_k})(D_x^{n-1} u^{m_k})^2 + \right.$$

$$+2c_n(u^{m_k} D_x^{n-1} u^{m_k}) P_{m_k}^{\perp} \left[ D_x \sum_{i=0}^{n-2} C_{n-2}^i D_x^{n-2-i} u^{m_k} D_x^{i+1} u^{m_k} \right] +$$

$$\left. +\sum_{i=0}^{n-2} \frac{\partial q_n(u^{m_k}, ..., D_x^{n-2} u^{m_k})}{\partial (D_x^i u^{m_k})} P_{m_k}^{\perp}[D_x^i(u^{m_k} u_x^{m_k})] \right\} dx = \int_0^A \left\{ c_n P_{m_k}^{\perp}(u^{m_k} u_x^{m_k})(D_x^{n-1} u^{m_k})^2 + \right.$$

$$+2c_n P_{m_k}^{\perp}(u^{m_k} D_x^{n-1} u^{m_k}) \left[ D_x \sum_{i=0}^{n-3} C_{n-2}^i D_x^{n-2-i} u^{m_k} D_x^{i+1} u^{m_k} \right] +$$

$$\left. +\sum_{i=0}^{n-2} \frac{\partial q_n(u^{m_k}, ..., D_x^{n-2} u^{m_k})}{\partial (D_x^i u^{m_k})} P_{m_k}^{\perp}[D_x^i(u^{m_k} u_x^{m_k})] \right\} dx.$$

These equalities imply the statement of Lemma IV.3.5.□

Lemma IV.3.6 *For any integer $n \geq 2$ and any segment $I = [t_0 - T, t_0 + T]$ there exists a function $\beta_n(s)$ continuous and nondecreasing on $[0, +\infty)$ such that for any $u_0 \in H^n_{per}(A)$ the quantities $\|u^m(\cdot, t)\|_n$ are bounded by $\beta_n(\|u_0\|_n)$ uniformly in $t \in I$ and $m = 1, 2, 3, ...$ (here $u^m(\cdot, t)$ are solutions of the problem (IV.3.4),(IV.3.5)).*

Proof in view of Lemma IV.3.5 follows from the estimate $(t > t_0)$:

$$\max_{t \in I} \sigma_n(\|u^m(\cdot, t)\|_{n-1}) \geq 2|E_n(u^m(\cdot, t)) - E_n(u^m(\cdot, t_0))| \geq$$

$$\geq |\, |D_x^n u^m(\cdot,t)|^2_{L_2(0,A)} - |D_x^n u^m(\cdot,t_0)|^2_{L_2(0,A)} \,| - \theta_n(\|u^m(\cdot,t)\|_{n-1}) - \theta_n(\|u^m(\cdot,t_0)\|_{n-1}),$$

where $\sigma_n(s)$ and $\theta_n(s)$ are functions continuous and nondecreasing on the half-line $[0, +\infty)$ $(n \geq 2)$. Indeed, fix an arbitrary $R > 0$ and let $\|u_0\|_n \leq R$. Then, taking into account that as earlier $\|u^m(\cdot,t)\|_1 \leq C(R)$ for all $t$, we get step by step:

$$\|u^m(\cdot,t)\|_2 \leq C_2(R), ..., \|u^m(\cdot,t)\|_n \leq C_n(R),$$

and Lemma IV.3.6 is proved.□

Lemma IV.3.7 *For any $u_0 \in H^n_{per}(A)$, $T > 0$ and integer $n \geq 2$ the following takes place*

$$\max_{t \in [t_0-T,t_0+T]} \|u^{m_k}(\cdot, t) - u(\cdot, t)\|_{n-1} \to 0 \quad \text{as} \quad k \to \infty.$$

Proof. First, let $u_0 \in H^\infty$. Using Lemmas IV.3.4 and IV.3.6 and Theorem IV.3.1, we get:

$$\frac{1}{2}\frac{d}{dt}\int_0^A |D_x^{n-1}(u^{m_k}(\cdot,t) - u(\cdot,t))|^2 dx =$$

$$= -\frac{1}{2}\int_0^A D_x^{n-1}(u^{m_k}(\cdot,t) - u(\cdot,t)) \times D_x^n[(u^{m_k}(\cdot,t))^2 - (u(\cdot,t))^2]dx +$$

$$+\frac{1}{2}\int_0^A P_{m_k}^\perp[D_x^{n-1}(u(\cdot,t))]D_x^n[(u^{m_k}(\cdot,t))^2]dx \leq$$

$$\leq C_1(\|u_0\|_n)\|P_{m_k}^\perp(u(\cdot,t))\|_n + C_2(\|u(\cdot,t_0)\|_n)\|u^{m_k}(\cdot,t) - u(\cdot,t)\|^2_{n-1},$$

where $\|P_{m_k}^\perp(u(\cdot,t))\|_n \to 0$ as $k \to \infty$ for any $t$ and $\|P_{m_k}^\perp(u(\cdot,t))\|_n \leq C$ for all $t$. By analogy

$$\frac{1}{2}\frac{d}{dt}|u^{m_k}(\cdot,t) - u(\cdot,t)|^2_{L_2(0,A)} \leq$$

$$\leq C_1(\|u_0\|_n)\|P^\perp_{m_k}(u(\cdot,t))\|_n + C_2(\|u_0\|_n)\|u^{m_k}(\cdot,t) - u(\cdot,t)\|^2_{n-1}.$$

Hence,

$$\|u^{m_k}(\cdot,t) - u(\cdot,t)\|^2_{n-1} \leq \|u^{m_k}(\cdot,t_0) - u(\cdot,t_0)\|^2_{n-1} + C_3\int_{t_0}^t \|u^{m_k}(\cdot,s) - u(\cdot,s)\|^2_{n-1} ds + a_{m_k},$$

where $a_{m_k} \to +0$ as $k \to \infty$. For an arbitrary $u_0 \in H^n_{per}(A)$ the latter inequality is also obviously valid. For $t < t_0$ similar estimates take place and, due to the Gronwell's lemma, Lemma IV.3.7 is proved. $\square$

<u>Lemma IV.3.8</u> *For any* $u_0 \in H^n_{per}(A)$ *and any* $t \in R$

$$\lim_{k\to\infty}[E_n(u^{m_k}(\cdot,t)) - E_n(u^{m_k}(\cdot,t_0))] = 0.$$

<u>Proof.</u> By Lemmas IV.3.5 and IV.3.6, since

$$|E_n(u^{m_k}(\cdot,t)) - E_n(u^{m_k}(\cdot,t_0))| \leq$$

$$\leq \left| \int_{t_0}^t \gamma_n(R, \max_{\substack{0\leq i,j\leq n-1 \\ \cdot +j\neq 2n-2}} |P^\perp_{m_k}(D^i_x u^{m_k}(\cdot,s) \times D^j_x u^{m_k}(\cdot,s))|_{L_2(0,A)} + \right.$$

$$\left. + \|P^\perp_{m_k}(u^{m_k} u^{m_k}_x)\|_1) ds \right|,$$

where $\|u^{m_k}(\cdot,s)\|_{n-1}$ are bounded uniformly with respect to $s \in [t_0,t]$ and positive integer $k$, it suffices to prove that

$$\lim_{k\to\infty}|P^\perp_{m_k}[D^i_x u^{m_k}(\cdot,s) \times D^j_x u^{m_k}(\cdot,s)]|_{L_2(0,A)} = \lim_{k\to\infty}\|P^\perp_{m_k}(u^{m_k}(\cdot,s)u^{m_k}_x(\cdot,s))\|_1 = 0$$

for all $s \in [t_0,t]$ and $0 \leq i,j \leq n-1$ where $i+j \neq 2n-2$. We have:

$$|P^\perp_{m_k}[D^i_x u^{m_k}(\cdot,s) \times D^j_x u^{m_k}(\cdot,s)]|_{L_2(0,A)} \leq |P^\perp_{m_k}[D^i_x u^{m_k}(\cdot,s) \times D^j_x u^{m_k}(\cdot,s) -$$

$$- D^i_x u(\cdot,s) \times D^j_x u(\cdot,s)]|_{L_2(0,A)} + |P^\perp_{m_k}[D^i_x u(\cdot,s) \times D^j_x u(\cdot,s)]|_{L_2(0,A)} \to 0$$

as $k \to \infty$ by Lemma IV 3.7 and Theorem IV.3.1. By analogy

$$\lim_{k\to\infty}\|P^\perp_{m_k}(u^{m_k}(\cdot,t)u^{m_k}_x(\cdot,t))\|_1 = 0$$

for any $t$, and Lemma IV.3.8 is proved. $\square$

Now just as in the proof of Theorem I.1.5, one can prove that for any $t$

$$\|u^{m_k}(\cdot,t) - u(\cdot,t)\|_n \to 0$$

as $k \to \infty$ if $u_0 \in H_{per}^n(A)$, and Proposition IV.3.3 is proved.$\square$

Corollary IV.3.9 *Let $n \geq 3$ be integer. Then, for any $u_0 \in H_{per}^{n-1}(A)$ and any $t$*
$$\lim_{m \to \infty} [E_n(u^m(\cdot, t)) - E_n(u^m(\cdot, t_0))] = 0.$$
Proof *repeats the proof of Lemma IV.3.8 in view of the proved Proposition IV.3.3.*$\square$

Below, to prove Theorem IV.3.2, we shall also need three statements.

Proposition IV.3.10 *For any $n \geq 2$, $u_0 \in H_{per}^n(A)$, $\epsilon > 0$ and $t \in R$ there exists $\delta > 0$ such that*
$$\|u^m(\cdot, t) - u_1^m(\cdot, t)\|_n < \epsilon$$
*for any $m = 1, 2, 3, ...$ and an arbitrary $L_m$-solution $u_1^m(\cdot, t)$ of equation (IV.3.4) satisfying*
$$\|u^m(\cdot, t_0) - u_1^m(\cdot, t_0)\|_n < \delta$$
*(here $u^m(x, t)$ is the solution of the problem (IV.3.4),(IV.3.5)).*

Proof. Suppose the contrary. Then there exists $\epsilon > 0$ such that for any $\delta > 0$ there are $m$ and $u_1 \in L_m$ satisfying
$$\|u_1^m(\cdot, t) - u^m(\cdot, t)\|_n \geq \epsilon \quad \text{and} \quad \|u_1^m(\cdot, t_0) - u^m(\cdot, t_0)\|_n < \delta$$
(here $u_1^m$ is a solution of the problem (IV.3.4),(IV.3.5) with $u_0 = u_1$). Then, there exist sequences $m_k$ and $u_1^{m_k} \in L_{m_k}$, where $\|u_1^{m_k} - P_{m_k} u_0\|_n \to 0$ as $k \to \infty$, such that
$$\|u_1^{m_k}(\cdot, t) - u^{m_k}(\cdot, t)\|_n \geq \epsilon,$$
where $u_1^{m_k}$ is the solution of the problem (IV.3.4),(IV.3.5) with $u_0 = u_1^{m_k}$ and $m = m_k$. Clearly, we can accept that the sequence $m_k$ is unbounded and, also, that $m_k \to +\infty$ as $k \to \infty$. But then by Proposition IV.3.3 $u_1^{m_k}(\cdot, t) \to u(\cdot, t)$ and $u^{m_k}(\cdot, t) \to u(\cdot, t)$ in $H_{per}^n(A)$ as $k \to \infty$, i. e. we get a contradiction. Proposition IV.3.10 is proved.$\square$

Proposition IV.3.11 *Let $n \geq 3$ be integer. Then, for any $t \in R$ there exists a function $\eta_n(s)$ nondecreasing and continuous on $[0, +\infty)$ such that*
$$|E_n(u^m(\cdot, t)) - E_n(u^m(\cdot, t_0))| \leq \eta_n(\|u^m(\cdot, t_0)\|_{n-1})$$
*for all $m = 1, 2, 3, ...$ and $u_0 \in H_{per}^{n-1}(A)$.*

Proof *follows from Lemmas IV.3.5 and IV.3.6.*$\square$

Proposition IV.3.12 *Let $n \geq 3$ be integer and let $K \subset H_{per}^{n-1}(A)$ be a compact set. Then for any $t \in R$   $E_n(u^m(\cdot, t)) - E_n(u^m(\cdot, t_0)) \to 0$ as $m \to \infty$ uniformly with respect to $u_0 \in K$ (here $u^m(\cdot, t_0) = P_m u_0$).*

Proof. We first prove that for any $\epsilon > 0$ and $\bar{u} \in K$ there exist $r > 0$ and a number $m_0$ such that

$$|E_n(u^m(\cdot, t)) - E_n(u^m(\cdot, t_0))| < \epsilon$$

for all $u_0 \in B_r(\bar{u}) = \{u \in H_{per}^{n-1}(A) : \|u - \bar{u}\|_{n-1} < r\}$ and all $m \geq m_0$. In view of Lemma IV.3.5, we have for any $r > 0$ and $u_0 \in B_r(\bar{u})$:

$$|E_n(u^m(\cdot, t)) - E_n(u^m(\cdot, t_0))| \leq \left| \int_{t_0}^{t} \gamma_n(R, \max_{\substack{0 \leq i,j \leq n-1 \\ i+j \neq 2n-2}} |P_m^\perp(D_x^i u^m(\cdot, s) \times D_x^j u^m(\cdot, s))|_{L_2(0,A)} + \right.$$

$$\left. + \|P_m^\perp(u^m(\cdot, s)u_x^m(\cdot, s))\|_1)ds \right|.$$

Let us estimate the integral from the right-hand side of the inequality. We have:

$$|P_m^\perp[D_x^i u^m(\cdot, s) \times D_x^j u^m(\cdot, s)]|_{L_2(0,A)} \leq$$

$$\leq |D_x^i u^m(\cdot, s) \times D_x^j u^m(\cdot, s) - D_x^i \bar{u}^m(\cdot, s) \times D_x^j \bar{u}^m(\cdot, s)|_{L_2(0,A)} +$$

$$+ |D_x^i \bar{u}^m(\cdot, s) \times D_x^j \bar{u}^m(\cdot, s) - D_x^i \bar{u}(\cdot, s) \times D_x^j \bar{u}(\cdot, s)|_{L_2(0,A)} +$$

$$+ |P_m^\perp[D_x^i \bar{u}(\cdot, s) \times D_x^j \bar{u}(\cdot, s)]|_{L_2(0,A)},$$

where $\bar{u}^m(\cdot, s)$ and $\bar{u}(\cdot, s)$ are the solutions of the problems (IV.3.4),(IV.3.5) and (IV.3.1)-(IV.3.3), respectively, with $u_0 = \bar{u}$.

In view of Proposition IV.3.3 and Theorem IV.3.1, the second and third terms in the right-hand side of this inequality tend to zero as $m \to \infty$ and do not depend on $u_0$. Further, by Proposition IV.3.10 for arbitrary $\epsilon > 0$ and $t \in R$ there exists $r > 0$ such that the first term in the right-hand side of the latter inequality is smaller than $\epsilon$ for all $u_0 \in B_r(\bar{u})$ and all $m$. Estimating in a similar way the term $\|P_m^\perp(u^m u_x^m)\|_1$, we get that for any $\epsilon > 0$ and $s$ there exist $r_0 > 0$ and a number $m_0$ such that

$$\psi(m,r) = \sup_{u_0 \in B_r(\bar{u})} \gamma_n(R, \max_{\substack{0 \leq i,j \leq n-1 \\ i+j \neq 2n-2}} |P_m^\perp[D_x^i u^m(\cdot, s) \times D_x^j u^m(\cdot, s)]|_{L_2(0,A)} +$$

$$+ \|P_m^\perp(u^m(\cdot, s)u_x^m(\cdot, s))\|_1) < \epsilon$$

for all $0 < r < r_0$ and $m \geq m_0$, if $u_0 \in B_r(\bar{u})$. In addition, the function $\psi(m,r)$ is bounded for all $0 < r < 1$ and $m = 1, 2, 3, \ldots$ by Lemma IV.3.6. This implies (with the use, for example, of the Egorov theorem) that for any $\epsilon > 0$ there exist $r \in (0, r_0)$ and a number $m_0 > 0$ such that

$$|E_n(u^m(\cdot, t)) - E_n(u^m(\cdot, t_0))| < \epsilon \qquad (\text{IV.3.6})$$

for all $m \geq m_0$ if $\|u_0 - \bar{u}\|_{n-1} < r$.

Fix an arbitrary $\epsilon > 0$ and compare to each point of the compact set $K$ a ball possessing the above property. Let $B_{r_1}(u_1), ..., B_{r_l}(u_l)$ be a finite covering of the set $K$ by these balls and let $m_1, ..., m_l$ be numbers such that for any $i$ the relation (IV.3.6) takes place if $m \geq m_i$ and $u_0 \in B_{r_i}(u_i)$. Then, obviously (IV.3.6) is also valid for all $u_0 \in K$ if $m \geq \max\limits_i m_i$, and Proposition IV.3.12 is proved.□

Fix an arbitrary integer $n \geq 3$. From here $\{e_k\}$ is the orthonormal basis in $H_{per}^{n-1}(A)$ consisting of eigenvectors of the operator $S$ of the above-indicated kind (i. e. $e_k$ are the above-defined functions with the change of $n$ by $n-1$ in the definition). Let also $\omega_i = (1 + \lambda_i^n)^{-1}(1 + \lambda_i^{n-1})$, where $i = 0, 1, 2, ....$ Then, $\omega_i$ are eigenvalues of $S$.

Consider in the subspaces $L_k \subset H_{per}^{n-1}(A)$ spanned over vectors $\{e_i\}_{i=0,1,2,...,2k}$ the finite-dimensional Gaussian measures $w_k$ defined by the rule

$$w_k(\Omega) = (2\pi)^{-\frac{2k+1}{2}} \prod_{i=0}^{2k} \omega_i^{-\frac{1}{2}} \int_F e^{-\frac{1}{2}\sum_{i=0}^{2k}\omega_i^{-1}z_i^2} dz_0...dz_{2k},$$

where $\Omega = \{u \in L_k | \, [(u, e_0)_{n-1}, ..., (u, e_{2k})_{n-1}] \in F\}$ and $F \subset R^{2k+1}$ are Borel sets. Then, $w_k$ is a Borel measure in $L_k$ for any $k$. According to results from Section 4.1, the measures $w_k$ can be considered as Borel measures in $H_{per}^{n-1}(A)$, and by Lemma IV.1.10 the sequence $\{w_k\}_{k=1,2,...}$ weakly converges to the infinite-dimensional Gaussian measure $w^n$. For a Borel set $\Omega \subset H_{per}^n(A)$ we also set

$$\mu_k(\Omega) = \int_\Omega e^{-J_n(u)} dw_k(u).$$

Since the functional $J_n$ is obviously continuous in $H_{per}^{n-1}(A)$ and bounded on bounded subsets of this space, the measures $\mu_k$ and $\mu^n$ are well-defined in $H_{per}^{n-1}(A)$. Let $h_m(u,t)$ be the dynamical system with the phase space $L_m$ generated by the system (IV.3.4),(IV.3.5) so that the function $h_m(\cdot, t)$ for any fixed $t$ transforms $L_m$ into $L_m$ and any $u_0 \in L_m$ maps into the solution $u^m(\cdot, t + t_0)$ of the problem (IV.3.4),(IV.3.5) taken at the moment of time $t + t_0$. Obviously $h_m(\cdot, t)$ for a fixed $t$ transforms also $H_{per}^{n-1}(A)$ into $L_m$ according to the rule $h_m(u,t) = h_m(P_m u, t)$.

<u>Lemma IV.3.13</u> Let $t \in R$ and $\Omega \subset W_2^{n-1}$ be a closed bounded set. Then

$$\lim_{m\to\infty} (\mu_m(h_m(\Omega, t)) - \mu_m(\Omega)) = 0.$$

<u>Proof.</u> Let us rewrite the system (IV.3.4),(IV.3.5) in the coordinates $z(t) = (z_0(t), ..., z_{2m}(t))$ where $u^m(x,t) = z_0(t)e_0(x) + ... + z_{2m}(t)e_{2m}(x)$. Then, we get

$$\dot{z}(t) = J\nabla_z H(z(t)),  \qquad (IV.3.7)$$

$$\dot{z}\cdot(t_0) = (u_0, e_i)_{n-1}, \quad i = 0, 1, ..., 2m, \qquad (IV.3.8)$$

where $H(z) = E_1(z_0 e_0 + ... + z_{2m} e_{2m})$ and $J$ is a skew-symmetric $(2m+1) \times (2m+1)$ matrix (i. e. $J^* = -J$), $(J)_{2k-1,2k} = -\frac{2\pi k}{A}\left(1 + \left(\frac{2\pi k}{A}\right)^{2n-2}\right) = -(J)_{2k,2k-1}$ ($k = 1, 2, ..., m$) and $(J)_{k,l} = 0$ for all other values of the indexes $k, l = 0, 1, ..., 2m$.

Let us prove that $\overline{D} = \left|\det\left(\frac{\partial z_i(t)}{\partial z_{0,j}}\right)_{i,j=\overline{0,2m}}\right| = 1$ for all $t$. Indeed, according to Theorem IV.1.3, the Lebesgue measure $\sigma_m(\Omega) = \int_\Omega dz_0...dz_{2m}$ is an invariant measure for the dynamical system with the phase space $L_m$ generated by the problem (IV.3.7),(IV.3.8). Therefore,

$$\sigma_m(h_m(\Omega, t)) = \int_{h_m(\Omega,t)} dz_0...dz_{2m} = \int_\Omega \overline{D} dz_0...dz_{2m} = \int_\Omega dz_0...dz_{2m}$$

for an arbitrary Borel set $\Omega \subset R^{2m+1}$. In view of the continuity of the function $\overline{D}$, this immediately implies that $\overline{D} \equiv 1$.

Let us take an arbitrary closed bounded set $\Omega \subset H_{per}^{n-1}(A)$. In view of the above arguments, we get:

$$\mu_m(h_m(\Omega, t)) = \int_\Omega e^{E_n(P_m u) - E_n(h_m(u,t))} d\mu_m(u).$$

Further,

$$|\mu_m(\Omega) - \mu_m(h_m(\Omega, t))| \leq \int_\Omega |1 - e^{E_n(P_m u) - E_n(h_m(u,t))}| d\mu_m(u),$$

therefore, according to Proposition IV.3.11 and Lemma IV.3.6, we obtain that the integrand in the right-hand side of this equality is a function bounded uniformly with respect to integer $m > 0$ and $u \in \Omega$. Take an arbitrary $\epsilon > 0$ and a compact set $K \subset H_{per}^{n-1}(A)$ such that $\mu(\Omega \setminus K) < \epsilon$, the existence of which can be proved as in the proof of Theorem IV.1.8. By Proposition IV.3.12,

$$\lim_{m\to\infty} [\mu_m(K \cap \Omega) - \mu_m(h_m(K \cap \Omega, t))] = 0,$$

hence, by Proposition IV.3.11, we get the relation

$$\limsup_{m\to\infty} [\mu_m(\Omega) - \mu_m(h_m(\Omega, t))] \leq C_1 \epsilon,$$

which, in view of the arbitrariness of $\epsilon > 0$, yields the statement of Lemma IV.3.13.□

Corollary IV.3.14 *For any bounded open set* $\Omega \subset H_{per}^{n-1}(A)$ *and for any* $t \in R$

$$\lim_{m\to\infty} |\mu_m(\Omega) - \mu_m(h_m(\Omega, t))| = 0.$$

Lemma IV.3.15 *Let* $\Omega \subset H_{\text{per}}^{n-1}(A)$ *be a bounded open set and* $t \in R$. *Then* $\mu^n(\Omega) = \mu^n(h^{n-1}(\Omega, t))$.

Proof. By Theorem IV.3.1 and Lemma IV.3.4 $h^{n-1}(\Omega, t)$ is a bounded open set in $H_{\text{per}}^{n-1}(A)$, too. Take an arbitrary $\epsilon > 0$. Then, there exists a compact set $K \subset \Omega$ such that $\mu^n(\Omega \setminus K) < \epsilon$. Let $K_1 = h^{n-1}(K, t)$. Then, $K_1$ is a compact set, too, and $K_1 \subset h^{n-1}(\Omega, t) = \Omega_1$. Let $\alpha = \min\{\text{dist}(K, \partial\Omega); \text{dist}(K_1, \partial\Omega_1)\}$, where $\text{dist}(A, B) = \inf_{u \in A, \ v \in B} \|u - v\|_{n-1}$ and $\partial A$ is a boundary of a set $A \subset H_{\text{per}}^{n-1}(A)$. Clearly, $\alpha > 0$. By Proposition IV.3.10 for any $u \in K$ there exists a ball $B_r(u) = \{v \in H_{\text{per}}^{n-1}(A) : \|u - v\|_{n-1} < r\}$ of a positive radius $r$ such that

$$\text{dist}(h_m(u, t); h_m(v, t)) < \frac{\alpha}{3}$$

for all $v \in B_r(u)$ and all $m$. Let $B_1, ..., B_l$ be a finite covering of the compact set $K$ by these balls. Let also $\Omega_\beta = \{v \in \Omega_1 : \text{dist}(v, \partial\Omega_1) \geq \beta\}$, where $\beta > 0$, and $B = \bigcup_{i=1}^{l} B_i$.

Since in view of Proposition IV.3.3 $h_m(u, t) \rightarrow h^{n-1}(u, t)$ in $H_{\text{per}}^{n-1}(A)$ as $m \rightarrow \infty$ for any $u \in H_{\text{per}}^{n-1}(A)$, we get that

$$h_m(B, t) \subset \Omega_{\frac{\alpha}{4}}$$

for all sufficiently large $m$. Further, by Lemma IV.1.11 and Corollary IV.3.14

$$\mu^n(\Omega) \leq \mu^n(B) + \epsilon \leq \liminf_{m \rightarrow \infty} \mu_m(B) + \epsilon = \liminf_{m \rightarrow \infty} \mu_m(h_m(B, t)) + \epsilon \leq \mu^n(\Omega_1) + \epsilon.$$

Therefore, in view of the arbitrariness of $\epsilon > 0$

$$\mu^n(\Omega) \leq \mu^n(\Omega_1).$$

By analogy $\mu^n(\Omega_1) \leq \mu^n(\Omega)$. Hence,

$$\mu^n(\Omega) = \mu^n(\Omega_1),$$

and Lemma IV.3.15 is proved.□

Let us prove Theorem IV.3.2. First, let $\Omega \subset H_{\text{per}}^{n-1}(A)$ be an open (generally unbounded) set. We set

$$\Omega^k = \{u \in \Omega : \|h^{n-1}(u, t)\|_{n-1} + \|u\|_{n-1} < k\},$$

where $k > 0$. Then $\Omega = \bigcup_{k=1}^{\infty} \Omega^k$, and each set $\Omega^k$ is open and bounded; in addition, $h^{n-1}(\Omega, t) = \bigcup_{k=1}^{\infty} h^{n-1}(\Omega^k, t)$ and $\mu^n(\Omega^k) = \mu^n(h^{n-1}(\Omega^k, t))$ by Lemma IV.3.15. Therefore,

$$\mu^n(h^{n-1}(\Omega, t)) = \lim_{k \rightarrow \infty} \mu^n(h^{n-1}(\Omega^k, t)) = \lim_{k \rightarrow \infty} \mu^n(\Omega^k) = \mu^n(\Omega).$$

Let now $A \subset H_{\mathrm{per}}^{n-1}(A)$ be an arbitrary Borel set.  By Proposition IV.1.1, $h^{n-1}(A, t)$ is a Borel subset of the space $H_{\mathrm{per}}^{n-1}(A)$. The equality $\mu^n(A) = \mu^n(h^{n-1}(A, t))$ now can be obtained by approximations of the set $A$ by open sets from outside. The last two statements of Theorem IV.3.2 follow from Lemma IV.3.4, the continuity of the functionals $E_0, ..., E_{n-1}$ in $H_{\mathrm{per}}^{n-1}(A)$ and from their boundedness on any bounded set from $H_{\mathrm{per}}^{n-1}(A)$. Thus, Theorem IV.3.2 is completely proved.□.

## 4.4   Additional remarks

First of all, we note that there are approaches to explain the recurrence properties of trajectories of dynamical systems generated by the KdVE and NLSE which are not based on the application of the Poincaré recurrence theorem (see for example [16,20,53,66]). Here we do not consider these investigations in detail.

Concerning invariant measures, there is a number of papers devoted to their construction for dynamical systems generated by nonlinear partial differential equations. Some of them are indicated earlier. One of the first results in this direction is obtained in [105] where an invariant measure is constructed for a NLSE. Similar measures (associated with the energy conservation law $E_1$ in the case of the KdVE and $E$ in the case of the NLSE) are considered, for example, in [4,11,18,19,24,30,55,65,67,86,107-109,111-113]. In [4,24,55], the invariance of these measures is not proved. However, the invariance in our sense easily follows from a result of the paper [4] for an abstract equation. In [30], an invariant measure is constructed for a cubic nonlinear wave equation. Unfortunately, some important details of the proof seem to be not completely satisfactory in this paper. In [67] and [108], invariant measures are constructed for a nonlinear wave equation with a weak nonlinearity. Methods exploited in these two papers are quite different. In [11], the author, besides others, more explicitly and carefully reestablishes an abstract construction from [109].

Another problem related to invariant measures is connected with the case of the NLSE with superlinear nonlinearities such as $f(x, |u|^2)u = \lambda|u|^p u$ where $p > 0$ and $\lambda$ are constants. In our paper [107], we investigate this case and obtain sufficient conditions for the measure $\mu$ similar to those from Theorem IV.2.2 to be nonzero and finite for any ball $B \subset X$. The obtained conditions are the following: $0 < \mu(B) < \infty$ for any ball $B \subset X$ if there exist $C > 0$ and $d_1 > 0$, $d_2 \in (0, 1)$ such that $-C(1+s^{d_1}) \leq f(x, s) \leq C(1 + s^{d_2})$ for all $x, s$. For $f(|u|^2) = \lambda|u|^p$ this implies: $p > 0$ if $\lambda < 0$ and $p \in (0, 2)$ for $\lambda > 0$. Unfortunately, in this paper an important question remains open about the well-posedness of the initial-boundary value problem (IV.2.1)-(IV.2.4) with $N = 1$ in this superlinear case with initial data from a space like $L_2$. The required result is obtained by J. Bourgain [16,17] who proved the well-posedness in a sense of this problem for an arbitrary $\lambda$. This allowed the author of this paper to construct an invariant measure for the one-dimensional NLSE with the power nonlinearity and to show its boundedness in the above sense for $p \in (0,5)$ (see [18]). A result in this direction for the cubic NLSE is also presented in the paper [65].

In the paper [67], an invariant measure, associated with a higher conservation law containing the square of the second derivative of the unknown function, is constructed for the sinh-Gordon equation. A result analogous to the result of Theorem IV.3.2 on the existence of an infinite sequence of invariant measures for the problem periodic in the spatial variable for the usual cubic NLSE

$$iu_t + u_{xx} + \lambda |u|^2 u = 0,$$

where $\lambda$ is a constant, is presented in the paper [112]. In this connection, the following question may be posed: are there invariant measures associated with the lowest conservation laws $E_0, E_1, E_2$ for the KdVE (or the NLSE)? For example, consider the conservation law $E_0$. In this case, our question is not answered yet. However, we can make some comments on it. One could observe that in Theorems IV.2.2 and IV.3.2 an invariant measure on the phase space $H^{n-1}$ corresponds to the $n$th conservation law. Therefore, we can hypothesize that to construct a required measure, we should prove the well-posedness of the corresponding evolution problem for the KdVE with initial data from a space similar to the Sobolev space $H^{-1}$ (or at least from $H^{-\frac{1}{2}-\epsilon}$, $\epsilon > 0$). Unfortunately, we do not know any results like that. In our opinion, this is one of the main difficulties in the way of constructing invariant measures corresponding to the above conservation law.

Finally, in the paper [113], the following Cauchy problem for the NLSE written in the real form

$$u_t^1 - u_{xx}^2 + V(x)u^2 + f(x,(u^1)^2 + (u^2)^2)u^2 = 0, \quad x,t \in R, \qquad \text{(IV.4.1)}$$

$$u_t^2 + u_{xx}^1 - V(x)u^1 - f(x,(u^1)^2 + (u^2)^2)u^1 = 0, \quad x,t \in R, \qquad \text{(IV.4.2)}$$

$$u^i(x,t_0) = u_0^i, \quad i = 1,2, \qquad \text{(IV.4.3)}$$

where $V(x)$ is a real-valued function of $x \in R$, is considered. In this paper, it is assumed that the function $f$ satisfies conditions similar to those introduced in Theorem IV.2.2. The main hypothesis on the potential $V$ is that this function is positive, tends to $+\infty$ as $|x| \to \infty$ and increases as $|x| \to \infty$ so rapidly that the eigenvalues $\nu_n$ of the operator $\left(-\frac{d^2}{dx^2} + V(x)\right)^{-1}$ satisfy the condition $\sum_n \nu_n < +\infty$. Under hypotheses like the above-described, in this paper we construct an invariant measure for a dynamical system generated by the problem (IV.4.1)-(IV.4.3) on the phase space $X = L_2(R) \otimes L_2(R)$ where $L_2(R)$ is the real space. In addition, in this paper some results on the boundedness of the measure under consideration are obtained in the supercritical case $f(x,s) = \lambda |s|^p$.

# Bibliography

[1] **A. Ambrosetti, P.H. Rabinowitz.** *Dual variational methods in critical point theory and applications*, J. Funct. Anal. **14**, 349-381 (1973).

[2] **I.V. Amirkhanov, E.P. Zhidkov.** *A sufficient condition of the existence of a particle-like solution for some field models*, JINR Commun., Part I: P5-80-479, Part II: P5-80-585, Dubna (1980) (in Russian).

[3] **I.V. Amirkhanov, E.P. Zhidkov, G.I. Makarenko.** *A sufficient condition for the existence of a particle-like solution of a nonlinear scalar field equation*, JINR Commun., P5-11705, Dubna (1978) (in Russian).

[4] **A.A. Arsen'ev.** *On invariant measures for classical dynamical systems with an infinite-dimensional phase space*, Matem. Sbornik **121**, 297-309 (1983) (in Russian).

[5] **N.K. Bary.** *On bases in Hilbert space*, Doklady Akad. Nauk SSSR **54**, No 5, 383-386 (1946) (in Russian).

[6] **N.K. Bary.** *Biorthogonal systems and bases in Hilbert space*, Moskov. Gos. Univ. Učenye Zapiski **148**, Matematika **4**, 69-107 (1951) (in Russian); *Math. Rev.* **14**, 289 (1953).

[7] **T.B. Benjamin.** *The stability of solitary waves*, Proc. Royal Soc. London **A328**, 153-183 (1972).

[8] **T.B. Benjamin.** *Lectures on nonlinear wave motion*, In: *"Nonlinear Wave Motion"*, A. Newell (ed.), Lect. Appl. Math. **15**, AMS, Providence, 1974, p. 3-48.

[9] **H. Berestycki, P.L. Lions.** *Existence d'ondes solitaires dans des problèmes non linéaires du type Klein-Gordon*, C.R. Acad. Sci. **AB288**, No 7, 395-398 (1979).

[10] **H. Berestycki, P.L. Lions, L.A. Peletier.** *An ODE approach to the existence of positive solutions for semilinear problems in $R^N$*, Indiana Univ. Math. J. **30**, No 1, 141-157 (1981).

[11] **B. Bidegaray.** *Invariant measures for some partial differential equations*, Physica **D82**, 340-364 (1995).

[12] **Ph. Blanchard, J. Stube, L. Vazques.** *On the stability of solitary waves for classical scalar fields*, Ann. Inst. H. Poincaré, Phys. Theor. **47**, No 3, 309-336 (1987).

[13] **N.N. Bogoliubov, N.M. Krylov.** *General measure theory in nonlinear mechanics*, In: "*Selected papers of N.N. Bogoliubov*", Kiev, Naukova Dumka, 1969, p. 411-463 (in Russian).

[14] **J.L. Bona.** *On the stability of solitary waves*, Proc. Royal Soc. London **A344**, No 1638, 363-374 (1975).

[15] **J.L. Bona, P.E. Souganidis, W.A. Strauss.** *Stability and instability of solitary waves of Korteweg–de Vries type*, Proc. Royal Soc. London **A411**, No 1841, 395-412 (1987).

[16] **J. Bourgain.** *Fourier transform restriction phenomena for certain lattice subsets and applications to nonlinear evolution equations*, Geom. Funct. Anal. **3**, No 2 (1993). Part 1: *Schrödinger equations*, 107-156; Part 2: *The KDV-equation*, 209-262.

[17] **J. Bourgain.** *Global Solutions of Nonlinear Schrödinger Equations*, AMS, Providence, 1999.

[18] **J. Bourgain.** *Periodic nonlinear Schrödinger equation and invariant measures*, Commun. Math. Phys. **166**, 1-26 (1994).

[19] **J. Bourgain.** *Invariant measures for the 2D-defocusing nonlinear Schrödinger equation*, Commun. Math. Phys. **176**, 421-445 (1996).

[20] **J. Bourgain.** *Quasi-periodic solutions of Hamiltonian perturbations of 2D linear Schrödinger equations*, Ann. of Math. **148**, No 2, 363-439 (1998).

[21] **T. Cazenave.** *Stable solutions of the logarithmic Schrödinger equation*, Nonlinear Anal.: Theory, Meth. & Appl. **7**, 1127-1140 (1983).

[22] **T. Cazenave, M.J. Esteban.** *On the stability of stationary states for nonlinear Schrödinger equations with an external magnetic field*, Mat. Aplic. Comput. **7**, No 3, 155-168 (1988).

[23] **T. Cazenave, P.L. Lions.** *Orbital stability of standing waves for some nonlinear Schrödinger equations*, Commun. Math. Phys. **85**, No 4, 549-561 (1982).

[24] **I.D. Chueshov.** *Equilibrium statistical solutions for dynamical systems with an infinite number of degrees of freedom*, Math. USSR Sbornik **58**, No 2, 397-406 (1987).

[25] **C.V. Coffman.** *Uniqueness of the ground state solution for $\Delta u - u + u^3 - 0$ and a variational characterization of other solutions*, Arch. Rat. Mech. Anal. **46**, 81-95 (1972).

[26] **A. Cohen, T. Kappeler.** *Nonuniqueness for solutions of the Korteweg–de Vries equation*, Trans. Amer. Math. Soc. **312**, No 2, 819-840 (1989).

[27] **Yu.L. Daletskii, S.V. Fomin.** *Measures and Differential Equations in Infinite-Dimensional Spaces*, Nauka, Moscow, 1983 (in Russian).

[28] **N. Dunford, J.T. Schwartz.** *Linear operators*, Part 2: *Spectral theory. Self Adjoint Operators in Hilbert Space*, Interscience Publ., New York – London, 1963.

[29] **A.V. Faminskii.** *The Cauchy problem for the Korteweg–de Vries equation and for its generalizations.* In: "Trudy Semin. imeni I.G. Petrovskogo" **13**, Moscow, Moscow State Univ. Publ, 1988, p. 56-105 (in Russian).

[30] **L. Friedlander.** *An invariant measure for the equation $u_{tt} - u_{xx} + u^3 = 0$*, Commun. Math. Phys. **98**, 1-16 (1985).

[31] **I.M. Gelfand.** *Sur un lemme de la théorie des espaces linéaires*, Kharkov, Zapiski Matem. Obshestva, Ser. 4, **13**, 35-40 (1936).

[32] **I.M. Gelfand, A.M. Yaglom.** *Integration in functional spaces and its applications in quantum physics*, Uspekhi Matem. Nauk **11**, No 1, 77-114 (1956) (in Russian).

[33] **A.V. Giber, A.B. Shabat.** *On the Cauchy problem for the nonlinear Schrödinger equation*, Different. Uravnenija **6**, No 1, 137-146 (1970) (in Russian).

[34] **B. Gidas, Ni Wei-Ming, L. Nirenberg.** *Symmetry and related properties via the maximum principle*, Commun. Math. Phys. **68**, No 3, 209-243 (1979).

[35] **B. Gidas, Ni Wei-Ming, L. Nirenberg.** *Symmetry of positive solutions of nonlinear elliptic equations in $R^N$*, In: "Mathematical Analysis and Applications, pt. 1", New York, e.a., 1981, p. 369-402.

[36] **D. Gilbarg, N.S. Trudinger.** *Elliptic partial differential equations of second order*, Springer, Berlin, 1983.

[37] **J. Ginibre, G. Velo.** *On a class of nonlinear Schrödinger equations. I. The Cauchy problem, general case*, J. Funct. Anal. **32**, 1-32 (1979).

[38] **R.T. Glassey.** *On the blowing up of solutions to the Cauchy problem for nonlinear Schrödinger equations*, J. Math. Phys. **18**, 1794-1797 (1977).

[39] **I.C. Gohberg, M.S. Krein.** *Introduction to the Theory of Linear non Self-Adjoint Operators*, Moscow, Nauka, 1965 (in Russian).

[40] **M. Grillakis, J. Shatah, W. Strauss.** *Stability theory of solitary waves in the presence of symmetry. I*, J. Funct. Anal. **74**, No 1, 160-197 (1987).

[41] **P.H. Halmos.** *Measure Theory*, New York, 1950.

[42] **D.B. Henri, J.F. Perez, W.F. Wreszinski.** *Stability theory for solitary wave solutions of scalar field equation*, Commun. Math. Phys. **85**, No 3, 351-361 (1982).

[43] **I.D. Iliev, E.Kh. Khristov, K.P. Kirchev.** *Spectral Methods in Soliton Equations*, Longman Publ., Pitman Monographs and Surveys in Pure and Applied Mathematics **73**, 1994.

[44] **T. Kato.** *On the Cauchy problem for the (generalized) Korteweg–de Vries equation*, In: *"Advances in Math. Suppl. Stud."* **8**, 93-128 (1983).

[45] **T. Kato.** *On nonlinear Schrödinger equations*, Ann. Inst. H. Poincaré, Phys. Theor. **46**, No 1, 113-129 (1987).

[46] **C.E. Kenig, G. Ponce, L. Vega.** *Well-posedness of the initial value problem for the Korteweg–de Vries equation*, J. Amer. Math. Soc. **4**, No 2, 323-347 (1991).

[47] **I.T. Kiguradze, B.L. Shekhter.** *On a boundary problem arising in the nonlinear field theory*. In: *"Itogy Nauki i Tekhniki"*, Sovrem. Probl. Matem. **30**, Moscow, Nauka, 1987, p. 183-201 (in Russian).

[48] **A.N. Kolmogorov, I.G. Petrovskii, N.S. Piskunov.** *An investigation of an equation of diffusion with the increase of the quantity of matter and its application to a biological problem*, Bull. Moscow Gos. Univ. **1**, No 6, 1-26 (1937) (in Russian).

[49] **V.P. Kotlyarov, E.Ya. Khruslov.** *Asymptotic solitons*. In: *"Functional and Numerical Methods in Mathematical Physics"*, Kiev, Naukova Dumka, 1988, p. 103-107 (in Russian).

[50] M.D. Kruskal, R.M. Miura, C.S. Gardner, N.Z. Zubusky. *Korteweg-de Vries equation and its generalizations. V. Uniqueness and nonexistence of polynomial conservation laws*, J. Math. Phys. 11, No 3, 952-960 (1970).

[51] N.S. Kruzhkov, A.V. Faminskii. *Generalized solutions of the Cauchy problem for the Korteweg-de Vries equation*, Matem. Sbornik 120, No 3, 396-425 (1983) (in Russian).

[52] N.M. Kryloff, N.N. Bogoluboff. *La Théorie Generale de la Mesure dans son Application à l'étude des Systèmes Dynamiques de la Mécanique non linéaire*, Ann. of Math. 38, No 2, 65-113 (1937).

[53] S.B. Kuksin. *On Long-Time Behavior Solutions of Nonlinear Wave Equations*, In: "Proceedings of the XIth International Congress of Mathematical Physics", Paris, July 18-23, 1994. International Press, Boston, 1995, p. 273-277.

[54] M.K. Kwong. *Uniqueness of positive solutions of $\Delta u - u + u^p = 0$ in $R^N$*, Arch. Rat. Mech. Anal. 105, No 3, 243-266 (1989).

[55] J.L. Lebowitz, H.A. Rose, E.R. Speer. *Statistical mechanics of the nonlinear Schrödinger equation*, J. Statist. Phys. 50, No 3-4, 657-687 (1988).

[56] P.L. Lions. *Symmetry and compactness in Sobolev spaces*, J. Funct. Anal. 49, 315-334 (1982).

[57] P.L. Lions. *The concentration – compactness method in the calculus of variations. The locally compact case*, Ann. Inst. H. Poincaré, Anal. non Lineaire 1, Part 1: 109-145; Part 2: 223-283 (1984).

[58] P.L. Lions. *The concentration – compactness principle in the calculus of variations. (The limit case.)* Revista Matem. Iberoamericana, 1, No 1: 145-201; No 2: 45-121 (1985).

[59] J.W. Macky. *A singular nonlinear boundary value problem*, Pacif. J. Math. 78, No 2, 375-383 (1978).

[60] V.G. Makhankov. *Soliton Phenomenology*, Kluwer Acad. Publ., Dordrecht, 1990.

[61] V.G. Makhankov, Yu.P. Rybakov, V.I. Sanyuk. *Localized non-topological structures: construction of solutions and stability problems*, Uspekhi Fiz. Nauk 164, No 2, 121-148 (1994) (in Russian).

[62] **A.P. Makhmudov.** *Fundamentals of nonlinear spectral analysis*, Baku, Azerbaijanian Gos. Univ. Publ., 1984 (in Russian).

[63] **A.P. Makhmudov.** *Completeness of eigenelements for some non-linear operator equations*, Doklady Akad. Nauk SSSR **263**, No 1, 23-27 (1982) (in Russian).

[64] **V.A. Marchenko.** *The Cauchy problem for the Korteweg–de Vries equation with non-vanishing initial data*, In: *"Integrability and kinetic equations for solitons"*, V.G. Baryakhtar, V.E. Zakharov and V.M. Chernousenko (eds.), Kiev, Naukova Dumka, 1990, p. 168-212 (in Russian).

[65] **H.P. McKean.** *Statistical mechanics of nonlinear wave equations (4): Cubic Schrödinger*, Commun. Math. Phys. **168**, No 3, 479-491 (1995).

[66] **H.P. McKean, E. Trubowitz.** *Hill's operator and hyperelliptic function theory in the presence of infinitely many branch points*, Commun. Pure Appl. Math. **29**, 143-226 (1976).

[67] **H. McKean, K. Vaninsky.** *Statistical mechanics of nonlinear wave equations*, In: *"Trends and Perspectives in Applied Mathematics"*, L. Sirovich (ed.). Springer-Verlag, New York, 1994, p. 239-264.

[68] **K. McLeod, J. Serrin.** *Uniqueness of positive radial solutions of $\Delta u + f(u) = 0$ in $R^n$*, Arch. Rat. Mech. Anal. **99**, 115-145 (1987).

[69] **S.A. Nasibov.** *On the stability, destruction, delaying and self-channelling of solutions of a nonlinear Schrödinger equation*, Doklady Akad. Nauk SSSR, **285**, No 4, 807-811 (1985) (in Russian).

[70] **S.A. Nasibov.** *On optimal constants in some Sobolev inequalities and their application to a nonlinear Schrödinger equation*, Doklady Akad. Nauk SSSR **307**, No 2, 538-542 (1989) (in Russian).

[71] **Z. Nehari.** *On a nonlinear differential equation arising in nuclear physics*, Proc. Royal Irish. Acad. **62A**, 117-135 (1963).

[72] **V. Nemytskii, V. Stepanov.** *Qualitative Theory of Differential Equations*, Moscow – Leningrad, 1949 (in Russian).

[73] **R.S. Palais.** *The symmetries of solitons*, Bull. Amer. Math. Soc. **34**, No 4, 339-403 (1997).

[74] **S.I. Pohozaev.** *Eigenfunctions of the equation $\Delta u + \lambda f(u) = 0$*, Doklady Akad. Nauk SSSR **165**, No 1, 36-39 (1965) (in Russian).

[75] **S.I. Pohozaev.** *On an approach to nonlinear equations*, Doklady Akad. Nauk SSSR **247**, No 6, 1327-1331 (1979) (in Russian).

[76] **S.I. Pohozaev.** *On a method of fibering for solving nonlinear boundary-value problems*, Trudy MIAN imeni V.A. Steklova **192**, 146-163 (1990) (in Russian).

[77] **P.H. Rabinowitz.** *Variational methods for nonlinear elliptic eigenvalue problems*, Indiana Univ. Math. J. **23**, No 8, 729-754 (1974).

[78] **P.H. Rabinowitz.** *Minimax Methods in Critical Point Theory with Applications to Differential Equations*, CBMS Regional Conf. Ser. in Math. 65, AMS, Providence, 1988.

[79] **M. Reed, B. Simon.** *Methods of Modern Mathematical Physics*, Acad. Press, New York – London, Part 1: 1972; Part 2: 1975.

[80] **R.D. Richtmyer.** *Principles of Advanced Mathematical Physics*, Part 1, Springer, 1978.

[81] **H.A. Rose, M.I. Weinstein.** *On the bound states of the nonlinear Schrödinger equation with a linear potential*, Physica **D30**, No 1-2, 207-218 (1988).

[82] **G.H. Ryder.** *Boundary value problems for a class of nonlinear differential equations*, Pacif. J. Math. **22**, No 3, 477-503 (1967).

[83] **G. Sansone.** *Su un'equazione differenziale non lineare della fisica nucleare*, In: *Symposia Math.* **6**, Acad. Press, London and New York, 1971, 3-139.

[84] **J. Shatah, W.A. Strauss.** *Instability of nonlinear bound states*, Commun. Math. Phys. **100**, 173-190 (1985).

[85] **A.V. Skorokhod.** *Integration in Hilbert Space*, Moscow, Nauka, 1975 (in Russian).

[86] **S.I. Sobolev.** *On an invariant measure for a nonlinear Schrödinger equation*, Trudy Petrozavodskogo Gos. Universiteta, Matematika, No 2, 113-124 (1995) (in Russian).

[87] **W.A. Strauss.** *Existence of solitary waves in higher dimensions*, Commun. Math. Phys. **55**, 149-162 (1977).

[88] **W.A. Strauss.** *Nonlinear Wave Equations*, CBMS Regional Conf. Ser. in Math. **73**, AMS, Providence, 1989.

[89] **L.A. Takhtajan, L.D. Faddeev.** *A Hamiltonian approach in soliton theory,* Moscow, Nauka, 1986 (in Russian).

[90] **M. Tsutsumi.** *Nonexistence of global solutions to the Cauchy problem for the damped nonlinear Schrödinger equations,* SIAM J. Math. Anal. **15**, No 2, 357-366 (1984).

[91] **Y. Tsutsumi.** $L_2$-*solutions for nonlinear Schrödinger equations,* Funkcial. Ekvac. **30**, 115-125 (1987).

[92] **M.I. Weinstein.** *Lyapunov stability of ground states of nonlinear dispersive evolution equations,* Commun. Pure Appl. Math. **39**, 51-68 (1986).

[93] **M.I. Weinstein.** *Existence and dynamic stability of solitary wave solutions of equations arising in long wave propagation,* Commun. Part. Diff. Equat. **12**, No 10, 1133-1173 (1987).

[94] **V.E. Zakharov, S.V. Manakov, S.P. Novikov, L.P. Pitaevskii.** *Soliton Theory. Method of Inverse Problem,* Moscow, Nauka, 1980 (in Russian).

[95] **E.P. Zhidkov, K.P. Kirchev.** *Stability of soliton solutions of some nonlinear equations of mathematical physics,* Soviet J. of Particles and Nuclei **16**, No 3, 259-279 (1985).

[96] **E.P. Zhidkov, V.P. Shirikov.** *On a boundary-value problem for ordinary differential equations of the second order,* J Vichislit. Matem. i Matem. Fiz. **4**, No 5, 804-816 (1964) (in Russian).

[97] **E.P. Zhidkov, V.P. Shirikov, I.V. Puzynin.** *Cauchy problem and a boundary-value problem for a nonlinear ordinary differential equation of the second order,* JINR Commun., No 2005, Dubna (1965) (in Russian).

[98] **E.P. Zhidkov, P.E. Zhidkov.** *An investigation of particle-like solutions in some models of the nonlinear physics,* JINR Commun., P5-12609, P5-12610, Dubna (1979) (in Russian).

[99] **P.E. Zhidkov.** *On the existence in $R^n$ of a positive solution of the Dirichlet problem for a nonlinear elliptic equation,* JINR Commun. Dubna, 5-82-69 (1992) (in Russian).

[100] **P.E. Zhidkov.** *On the stability of the soliton solution of the nonlinear Schrödinger equation,* Differentsial. Uravnenija **22**, No 6, 994-1004 (1986) (in Russian).

[101] **P.E. Zhidkov.** *Stability of solutions of the kind of the solitary wave for a generalized Korteweg-de Vries equation*, JINR Commun., P5-86-800, Dubna (1986) (in Russian).

[102] **P.E. Zhidkov.** *A Cauchy problem for a nonlinear Schrödinger equation*, JINR Commun., P5-87-373, Dubna (1987) (in Russian).

[103] **P.E. Zhidkov.** *On the solvability of the Cauchy problem and a stability of some solutions of a nonlinear Schrödinger equation*, Matem. Modelirovanie 1, No 10, 155-160 (1989) (in Russian).

[104] **P.E. Zhidkov.** *On the Cauchy problem for a generalized Korteweg-de Vries equation with periodic initial data*, Differentsial. Uravnenija 26, No 5, 823-829 (1990) (in Russian).

[105] **P.E. Zhidkov.** *On an invariant measure for a nonlinear Schrödinger equation*, Soviet Math. Dokl. 43, No 2, 431-434 (1991).

[106] **P.E. Zhidkov.** *Existence of solutions to the Cauchy problem and stability of kink-solutions of the nonlinear Schrödinger equation*, Siberian Math. J. 33, No 2, 239-246 (1992).

[107] **P.E. Zhidkov.** *An invariant measure for the nonlinear Schrödinger equation*, JINR Commun., P5-94-199, Dubna (1994) (in Russian).

[108] **P.E. Zhidkov.** *An invariant measure for a nonlinear wave equation*, Nonlinear Anal.: Theory, Meth. & Appl. 22, 319-325 (1994).

[109] **P.E. Zhidkov.** *On invariant measures for some infinite-dimensional dynamical systems*, Ann. Inst. H. Poincaré, Phys. Theor. 62, No 3, 267-287 (1995).

[110] **P.E. Zhidkov, V. Zh. Sakbaev.** *On a nonlinear ordinary differential equation*, Matem. Zametki 55, No 4, 25-34 (1994) (in Russian).

[111] **P.E. Zhidkov.** *Invariant measures generated by higher conservation laws for the Korteweg-de Vries equation*, Sbornik: Mathematics 187, No 6, 803-822 (1996).

[112] **P.E. Zhidkov.** *On an infinite series of invariant measures for the cubic nonlinear Schrödinger equation*, Prepr. JINR, E5-96-77, Dubna (1996).

[113] **P.E. Zhidkov.** *Invariant measures for infinite-dimensional dynamical systems with applications to a nonlinear Schrödinger equation*, In: *"Algebraic and Geometric Methods in Mathematical Physics"*, A. Boutet de Monvel and V. Marchenko (eds.), Kluwer Acad. Publ., 1996, p. 471-476.

[114] **P.E. Zhidkov.** *Completeness of systems of eigenfunctions for the Sturm-Liouville operator with potential depending on the spectral parameter and for one non-linear problem*, Sbornik: Mathematics **188**, No 7, 1071-1084 (1997).

[115] **P.E. Zhidkov.** *Eigenfunction expansions associated with a nonlinear Schrödinger equation*, JINR Commun., E5-98-61, Dubna (1998).

[116] **P.E. Zhidkov.** *On the property of being a Riesz basis for the system of eigenfunctions of a nonlinear Sturm-Liouville-type problem*, Sbornik: Mathematics, **191**, No 3, 43-52 (2000) (in Russian).

[117] **P.E. Zhidkov.** *Basis properties of eigenfunctions of nonlinear Sturm – Liouville problems*, Electronic J. of Differential Equations **2000**, No 28, 1-13 (2000).

[118] **P.E. Zhidkov.** *On the property of being a basis for a denumerable set of solutions of a nonlinear Schrödinger-type boundary-value problem*, Nonlinear Anal.: Theory, Meth. & Appl. **43**, No 4, 471-483 (2001).

[119] **P.E. Zhidkov.** *Eigenfunction expansions associated with a nonlinear Schrödinger equation on a half-line*, Prepr. JINR, E5-99-144, Dubna (1999).

# Index

Printing: Weihert-Druck GmbH, Darmstadt
Binding: Buchbinderei Schäffer, Grünstadt

# Lecture Notes in Mathematics

For information about Vols. 1–1570
please contact your bookseller or Springer-Verlag

Vol. 1613: J. Azéma, M. Emery, P. A. Meyer, M. Yor (Eds.), Séminaire de Probabilités XXIX. VI, 326 pages. 1995.

Vol. 1614: A. Koshelev, Regularity Problem for Quasilinear Elliptic and Parabolic Systems. XXI, 255 pages. 1995.

Vol. 1615: D. B. Massey, Le Cycles and Hypersurface Singularities. XI, 131 pages. 1995.

Vol. 1616: I. Moerdijk, Classifying Spaces and Classifying Topoi. VII, 94 pages. 1995.

Vol. 1617: V. Yurinsky, Sums and Gaussian Vectors. XI, 305 pages. 1995.

Vol. 1618: G. Pisier, Similarity Problems and Completely Bounded Maps. VII, 156 pages. 1996.

Vol. 1619: E. Landvogt, A Compactification of the Bruhat-Tits Building. VII, 152 pages. 1996.

Vol. 1620: R. Donagi, B. Dubrovin, E. Frenkel, E. Previato, Integrable Systems and Quantum Groups. Montecatini Terme, 1993. Editors:M. Francaviglia, S. Greco. VIII, 488 pages. 1996.

Vol. 1621: H. Bass, M. V. Otero-Espinar, D. N. Rockmore, C. P. L. Tresser, Cyclic Renormalization and Auto-morphism Groups of Rooted Trees. XXI, 136 pages. 1996.

Vol. 1622: E. D. Farjoun, Cellular Spaces, Null Spaces and Homotopy Localization. XIV, 199 pages. 1996.

Vol. 1623: H.P. Yap, Total Colourings of Graphs. VIII, 131 pages. 1996.

Vol. 1624: V. Brınzanescu, Holomorphic Vector Bundles over Compact Complex Surfaces. X, 170 pages. 1996.

Vol.1625: S. Lang, Topics in Cohomology of Groups. VII, 226 pages. 1996.

Vol. 1626: J. Azéma, M. Emery, M. Yor (Eds.), Séminaire de Probabilités XXX. VIII, 382 pages. 1996.

Vol. 1627: C. Graham, Th. G. Kurtz, S. Méléard, Ph. E. Protter, M. Pulvirenti, D. Talay, Probabilistic Models for Nonlinear Partial Differential Equations. Montecatini Terme, 1995. Editors: D. Talay, L. Tubaro. X, 301 pages. 1996.

Vol. 1628: P.-H. Zieschang, An Algebraic Approach to Association Schemes. XII, 189 pages. 1996.

Vol. 1629: J. D. Moore, Lectures on Seiberg-Witten Invariants. VII, 105 pages. 1996.

Vol. 1630: D. Neuenschwander, Probabilities on the Heisenberg Group: Limit Theorems and Brownian Motion. VIII, 139 pages. 1996.

Vol. 1631: K. Nishioka, Mahler Functions and Transcendence.VIII, 185 pages.1996.

Vol. 1632: A. Kushkuley, Z. Balanov, Geometric Methods in Degree Theory for Equivariant Maps. VII, 136 pages. 1996.

Vol.1633: H. Aikawa, M. Essén, Potential Theory – Selected Topics. IX, 200 pages.1996.

Vol. 1634: J. Xu, Flat Covers of Modules. IX, 161 pages. 1996.

Vol. 1635: E. Hebey, Sobolev Spaces on Riemannian Manifolds. X, 116 pages. 1996.

Vol. 1636: M. A. Marshall, Spaces of Orderings and Abstract Real Spectra. VI, 190 pages. 1996.

Vol. 1637: B. Hunt, The Geometry of some special Arithmetic Quotients. XIII, 332 pages. 1996.

Vol. 1638: P. Vanhaecke, Integrable Systems in the realm of Algebraic Geometry. VIII, 218 pages. 1996.

Vol. 1639: K. Dekimpe, Almost-Bieberbach Groups: Affine and Polynomial Structures. X, 259 pages. 1996.

Vol. 1640: G. Boillat, C. M. Dafermos, P. D. Lax, T. P. Liu, Recent Mathematical Methods in Nonlinear Wave Propagation. Montecatini Terme, 1994. Editor: T. Ruggeri. VII, 142 pages. 1996.

Vol. 1641: P. Abramenko, Twin Buildings and Applications to S-Arithmetic Groups. IX, 123 pages. 1996.

Vol. 1642: M. Puschnigg, Asymptotic Cyclic Cohomology. XXII, 138 pages. 1996.

Vol. 1643: J. Richter-Gebert, Realization Spaces of Polytopes. XI, 187 pages. 1996.

Vol. 1644: A. Adler, S. Ramanan, Moduli of Abelian Varieties. VI, 196 pages. 1996.

Vol. 1645: H. W. Broer, G. B. Huitema. M. B. Sevryuk, Quasi-Periodic Motions in Families of Dynamical Systems. XI, 195 pages. 1996.

Vol. 1646: J.-P. Demailly, T. Peternell. G. Tian, A. N. Tyurin, Transcendental Methods in Algebraic Geometry. Cetraro, 1994. Editors: F. Catanese, C. Ciliberto. VII, 257 pages. 1996.

Vol. 1647: D. Dias, P. Le Barz. Configuration Spaces over Hilbert Schemes and Applications. VII. 143 pages. 1996.

Vol. 1648: R. Dobrushin, P. Groeneboom, M. Ledoux, Lectures on Probability Theory and Statistics. Editor: P. Bernard. VIII, 300 pages. 1996.

Vol. 1649: S. Kumar, G. Laumon, U. Stuhler, Vector Bundles on Curves – New Directions. Cetraro, 1995. Editor: M. S. Narasimhan. VII, 193 pages. 1997.

Vol. 1650: J. Wildeshaus, Realizations of Polylogarithms. XI, 343 pages. 1997.

Vol. 1651: M. Drmota, R. F. Tichy, Sequences, Discrepancies and Applications. XIII, 503 pages. 1997.

Vol. 1652: S. Todorcevic, Topics in Topology. VIII. 153 pages. 1997.

Vol. 1653: R. Benedetti, C. Petronio, Branched Standard Spines of 3-manifolds. VIII, 132 pages. 1997.

Vol. 1654: R. W. Ghrist, P. J. Holmes, M. C. Sullivan. Knots and Links in Three-Dimensional Flows. X, 208 pages. 1997.

Vol. 1655: J. Azéma. M. Emery, M. Yor (Eds.), Séminaire de Probabilités XXXI. VIII, 329 pages. 1997.

Vol. 1656: B. Biais, T. Björk, J. Cvitanic, N. El Karoui, E. Jouini, J. C. Rochet, Financial Mathematics. Bressanone, 1996. Editor: W. J. Runggaldier. VII, 316 pages. 1997.

Vol. 1657: H. Reimann. The semi-simple zeta function of quaternionic Shimura varieties. IX, 143 pages. 1997.

Vol. 1658: A. Pumarino. J. A. Rodrıguez, Coexistence and Persistence of Strange Attractors. VIII, 195 pages. 1997.

Vol. 1659: V, Kozlov, V. Maz`ya, Theory of a Higher-Order Sturm-Liouville Equation. XI, 140 pages. 1997.

Vol. 1660: M. Bardi. M. G. Crandall, L. C. Evans, H. M. Soner, P. E. Souganidis, Viscosity Solutions and Applications. Montecatini Terme, 1995. Editors: I. Capuzzo Dolcetta, P. L. Lions. IX, 259 pages. 1997.

Vol. 1661: A. Tralle, J. Oprea. Symplectic Manifolds with no Kähler Structure. VIII, 207 pages. 1997.

Vol. 1662: J. W. Rutter, Spaces of Homotopy Self-Equivalences – A Survey. IX, 170 pages. 1997.

Vol. 1663: Y. E. Karpeshina: Perturbation Theory for the Schrödinger Operator with a Periodic Potential. VII, 352 pages. 1997.

Vol. 1715: N. V. Krylov, M. Röckner, J. Zabczyk, Stochastic PDE's and Kolmogorov Equations in Infinite Dimensions. Cetraro, 1998. Editor: G. Da Prato. VIII, 239 pages. 1999.

Vol. 1716: J. Coates, R. Greenberg, K. A. Ribet, K. Rubin, Arithmetic Theory of Elliptic Curves. Cetraro, 1997. Editor: C. Viola. VIII, 260 pages. 1999.

Vol. 1717: J. Bertoin, F. Martinelli, Y. Peres, Lectures on Probability Theory and Statistics. Saint-Flour, 1997. Editor: P. Bernard. IX, 291 pages. 1999.

Vol. 1718: A. Eberle, Uniqueness and Non-Uniqueness of Semigroups Generated by Singular Diffusion Operators. VIII, 262 pages. 1999.

Vol. 1719: K. R. Meyer, Periodic Solutions of the N-Body Problem. IX, 144 pages. 1999.

Vol. 1720: D. Elworthy, Y. Le Jan, X-M. Li, On the Geometry of Diffusion Operators and Stochastic Flows. IV, 118 pages. 1999.

Vol. 1721: A. Iarrobino, V. Kanev, Power Sums, Gorenstein Algebras, and Determinantal Loci. XXVII, 345 pages. 1999.

Vol. 1722: R. McCutcheon, Elemental Methods in Ergodic Ramsey Theory. VI, 160 pages. 1999.

Vol. 1723: J. P. Croisille, C. Lebeau, Diffraction by an Immersed Elastic Wedge. VI, 134 pages. 1999.

Vol. 1724: V. N. Kolokoltsov, Semiclassical Analysis for Diffusions and Stochastic Processes. VIII, 347 pages. 2000.

Vol. 1725: D. A. Wolf-Gladrow, Lattice-Gas Cellular Automata and Lattice Boltzmann Models. IX, 308 pages. 2000.

Vol. 1726: V. Marić, Regular Variation and Differential Equations. X, 127 pages. 2000.

Vol. 1727: P. Kravanja, M. Van Barel, Computing the Zeros of Analytic Functions. VII, 111 pages. 2000.

Vol. 1728: K. Gatermann, Computer Algebra Methods for Equivariant Dynamical Systems. XV, 153 pages. 2000.

Vol. 1729: J. Azéma, M. Émery, M. Ledoux, M. Yor, Séminaire de Probabilités XXXIV. VI, 431 pages. 2000.

Vol. 1730: S. Graf, H. Luschgy, Foundations of Quantization for Probability Distributions. X, 230 pages. 2000.

Vol. 1731: T. Hsu, Quilts: Central Extensions, Braid Actions, and Finite Groups,. XII, 185 pages. 2000.

Vol. 1732: K. Keller, Invariant Factors, Julia Equivalences and the (Abstract) Mandelbrot Set. X, 206 pages. 2000.

Vol. 1733: K. Ritter, Average-Case Analysis of Numerical Problems. IX, 254 pages. 2000.

Vol. 1734: M. Espedal, A. Fasano, A. Mikelić, Filtration in Porous Media and Industrial Applications. Cetraro 1998. Editor: A. Fasano. 2000.

Vol. 1735: D. Yafaev, Scattering Theory: Some Old and New Problems. XVI, 169 pages. 2000.

Vol. 1736: B. O. Turesson, Nonlinear Potential Theory and Weighted Sobolev Spaces. XIV, 173 pages. 2000.

Vol. 1737: S. Wakabayashi, Classical Microlocal Analysis in the Space of Hyperfunctions. VIII, 367 pages. 2000.

Vol. 1738: M. Émery, A. Nemirovski, D. Voiculescu, Lectures on Probability Theory and Statistics. XI, 356 pages. 2000.

Vol. 1739: R. Burkard, P. Deuflhard, A. Jameson, J.-L. Lions, G. Strang, Computational Mathematics Driven by Industrial Problems. Martina Franca, 1999. Editors: V. Capasso, H. Engl, J. Periaux. VII, 418 pages. 2000.

Vol. 1740: B. Kawohl, O. Pironneau, L. Tartar, J.-P. Zolesio, Optimal Shape Design. Tróia, Portugal 1999. Editors: A. Cellina, A. Ornelas. IX, 388 pages. 2000.

Vol. 1741: E. Lombardi, Oscillatory Integrals and Phenomena Beyond all Algebraic Orders. XV, 413 pages. 2000.

Vol. 1742: A. Unterberger, Quantization and Non-holomorphic Modular Forms. VIII, 253 pages. 2000.

Vol. 1743: L. Habermann, Riemannian Metrics of Constant Mass and Moduli Spaces of Conformal Structures. XII, 116 pages. 2000.

Vol. 1744: M. Kunze, Non-Smooth Dynamical Systems. X, 228 pages. 2000.

Vol. 1745: V. D. Milman, G. Schechtman, Geometric Aspects of Functional Analysis. VIII, 289 pages. 2000.

Vol. 1746: A. Degtyarev, I. Itenberg, V. Kharlamov, Real Enriques Surfaces. XVI, 259 pages. 2000.

Vol. 1747: L. W. Christensen, Gorenstein Dimensions. VIII, 204 pages. 2000.

Vol. 1748: M. Růžička, Electrorheological Fluids: Modeling and Mathematical Theory. XV. 176 pages. 2001.

Vol. 1749: M. Fuchs, G. Seregin, Variational Methods for Problems from Plasticity Theory and for Generalized Newtonian Fluids. VI, 269 pages. 2001.

Vol. 1750: B. Conrad, Grothendieck Duality and Base Change. X, 296 pages. 2001.

Vol. 1751: N. J. Cutland, Loeb Measures in Practice: Recent Advances. XI, 111 pages. 2001.

Vol. 1752: Y. V. Nesterenko, P. Philippon, Introduction to Algebraic Independence Theory. XIII, 256 pages. 2001.

Vol. 1753: A. I. Bobenko, U. Eitner, Painlevé Equations in the Differential Geometry of Surfaces. VI, 120 pages. 2001.

Vol. 1754: W. Bertram, The Geometry of Jordan and Lie Structures. XVI, 269 pages. 2001.

Vol. 1755: J. Azéma, M. Émery, M. Ledoux, M. Yor, Séminaire de Probabilités XXXV. VI, 427 pages. 2001.

Vol. 1756: P. E. Zhidkov, Korteweg de Vries and Nonlinear Schrödinger Equations: Qualitative Theory. VII, 147 pages. 2001.

Vol. 1757: R. R. Phelps, Lectures on Choquet's Theorem. VII, 124 pages. 2001.

Vol. 1758: N. Monod, Continous Bounded Cohomology of Locally Compact Groups. X, 214 pages. 2001.

Vol. 1759: Y. Abe, K. Kopfermann, Toroidal Groups. VIII, 133 pages. 2001.

Vol. 1760: D. Filipović, Consistency Problems for Heath-Jarrow-Morton Interest Rate Models. VIII, 134 pages. 2001.

## Recent Reprints and New Editions

Vol. 1200: V. D. Milman, G. Schechtmann, Asymptotic Theory of Finite Dimensional Normed Spaces - Corrected Second Printing 2001. X, 156 pages. 1986.

Vol. 1618: G. Pisier, Similarity Problems and Completely Bounded Maps - Second, Expanded Edition VII, 198 pages. 2001.

Vol. 1629: J. D. Moore, Lectures on Seiberg-Witten Invariants - Second Edition. VIII, 121 pages. 2001.

Vol. 1702: J. Ma, J. Yong, Forward-Backward Stochastic Differential Equations and Their Applications - Corrected Second Printing 2000. XIII, 270 pages. 1999.

4. Lecture Notes are printed by photo-offset from the master-copy delivered in camera-ready form by the authors. Springer-Verlag provides technical instructions for the preparation of manuscripts. Macro packages in $T_EX$, $L^AT_EX2e$, $L^AT_EX2.09$ are available from Springer's web-pages at

http://www.springer.de/math/authors/b-tex.html.

Careful preparation of the manuscripts will help keep production time short and ensure satisfactory appearance of the finished book.

The actual production of a Lecture Notes volume takes approximately 12 weeks.

5. Authors receive a total of 50 free copies of their volume, but no royalties. They are entitled to a discount of 33.3 % on the price of Springer books purchase for their personal use, if ordering directly from Springer-Verlag.

Commitment to publish is made by letter of intent rather than by signing a formal contract. Springer-Verlag secures the copyright for each volume. Authors are free to reuse material contained in their LNM volumes in later publications: A brief written (or e-mail) request for formal permission is sufficient.

**Addresses:**

Professor J.-M. Morel
CMLA, Ecole Normale Supérieure de Cachan
61 Avenue du Président Wilson
94235 Cachan Cedex France
E-mail: Jean-Michel.Morel@cmla.ens-cachan.fr

Professor B. Teissier
Université Paris 7
UFR de Mathématiques
Equipe Géométrie et Dynamique
Case 7012
2 place Jussieu
75251 Paris Cedex 05
E-mail: Teissier@ens.fr

Professor F. Takens, Mathematisch Instituut,
Rijksuniversiteit Groningen, Postbus 800,
9700 AV Groningen, The Netherlands
E-mail: F.Takens@math.rug.nl

Springer-Verlag, Mathematics Editorial, Tiergartenstr. 17
D-69121 Heidelberg, Germany
Tel.: *49 (6221) 487-701
Fax: *49 (6221) 487-355
E-mail: lnm@Springer.de